T0341294

# Pearl Millet

# Pearl Millet

## Properties, Functionality and Its Applications

Edited by

### Sneh Punia Ph.D.,

*Department of Food Science and Technology, Chaudhary Devi Lal University, Sirsa (Haryana), India*

### Anil Kumar Siroha Ph.D.,

*Department of Food Science and Technology, Chaudhary Devi Lal University, Sirsa (Haryana), India*

### Kawaljit Singh Sandhu Ph.D.,

*Department of Food Science and Technology, Maharaja Ranjit Singh Punjab Technical University, Bathinda (Punjab), India*

### Suresh Kumar Gahlawat Ph.D.,

*Department of Biotechnology, Chaudhary Devi Lal University, Sirsa (Haryana), India*

and

### Maninder Kaur Ph.D.,

*Department of Food Science and Technology, Guru Nanak Dev University, Amritsar (Punjab), India*

CRC Press
Taylor & Francis Group
Boca Raton London New York

CRC Press is an imprint of the
Taylor & Francis Group, an **informa** business

CRC Press
Taylor & Francis Group
6000 Broken Sound Parkway NW, Suite 300
Boca Raton, FL 33487

© 2020 by Taylor & Francis Group, LLC

CRC Press is an imprint of Taylor & Francis Group, an Informa business

No claim to original U.S. Government works

Printed on acid-free paper

International Standard Book Number-13: 978-0-367-35486-2 (Hardback)

This book contains information obtained from authentic and highly regarded sources. Reasonable efforts have been made to publish reliable data and information, but the author and publisher cannot assume responsibility for the validity of all materials or the consequences of their use. The authors and publishers have attempted to trace the copyright holders of all material reproduced in this publication and apologize to copyright holders if permission to publish in this form has not been obtained. If any copyright material has not been acknowledged please write and let us know so we may rectify in any future reprint.

Except as permitted under U.S. Copyright Law, no part of this book may be reprinted, reproduced, transmitted, or utilized in any form by any electronic, mechanical, or other means, now known or hereafter invented, including photocopying, microfilming, and recording, or in any information storage or retrieval system, without written permission from the publishers.

For permission to photocopy or use material electronically from this work, please access www.copyright.com (http://www.copyright.com/) or contact the Copyright Clearance Center, Inc. (CCC), 222 Rosewood Drive, Danvers, MA 01923, 978-750-8400. CCC is a not-for-profit organization that provides licenses and registration for a variety of users. For organizations that have been granted a photocopy license by the CCC, a separate system of payment has been arranged.

**Trademark Notice:** Product or corporate names may be trademarks or registered trademarks, and are used only for identification and explanation without intent to infringe.

**Visit the Taylor & Francis Web site at
http://www.taylorandfrancis.com**

**and the CRC Press Web site at
http://www.crcpress.com**

# Contents

# About the Editors

**Sneh Punia, Ph.D.** is presently working as an Assistant Professor in the Department of Food Science and Technology, CDLU, Sirsa. Her area of interest includes antioxidants, starch, and the development of new products. She has published more than 20 research papers in national and international journals, and has presented more than 25 research papers to various national and international conferences. She also serves as a reviewer for various international journals.

**Anil Kumar Siroha, Ph.D.** is presently working as an Assistant Professor in the Department of Food Science and Technology, CDLU, Sirsa. His area of interest includes starch, starch modification, and development of new products. He has published more than 15 research papers in national and international journals. He is an active member of the Association of Food Scientists and Technologists (AFSTI), Mysore, India.

**Kawaljit Singh Sandhu, Ph.D.** is presently working as an Associate Professor in the Department of Food Science and Technology, Maharaja Ranjeet Singh Punjab Technical University, Bathinda. He was awarded a postdoctoral fellowship from Korea University, Seoul, South Korea. With more than ten years of research and teaching experience, his research interest is focused on starch, starch modification, antioxidants, nanotechnology, and drug delivery. Dr. Sandhu has carried out several research projects from UGC, New Delhi and DST, New Delhi. In 2008, he was given the Young Scientist Award from the Association of Food Scientists and Technologists (AFSTI), Mysore, India. Dr. Sandhu is an active member of AFSTI, India, the Association of Microbiologists of India, and the Korean Society of Food Science and Technology, South Korea. He has published more than 50 research papers in various national and international journals.

**Suresh Kumar Gahlawat, Ph.D.** is a Professor in the Department of Biotechnology, and Former Dean, Research, Chaudhary Devi Lal University (CDLU), Sirsa, India. He received a postdoctoral BOYSCAST fellowship and DBT Overseas Associateship from the Ministry of Science & Technology, Government of India, for carrying out research at FRS Marine Laboratory, Aberdeen, UK. He has completed four R&D projects from UGC, ICAR, and the Government of Haryana. His research interests include the development of molecular diagnostic methods for bacterial and viral diseases. He has published more than 70 research papers in journals of national and international repute, authored more than ten books, and supervised the M.Phil. and Ph.D. research work of 12 students. He is an active member of various international scientific organizations and societies, including the Association of Microbiologists of India.

**Maninder Kaur, Ph.D.** is Assistant Professor at the Department of Food Science and Technology, Guru Nanak Dev University, Amritsar, India. She has published more than 45 research articles in various international and national peer-reviewed journals. Dr. Kaur is a member of various scientific societies and associations, including being an active member of the Association of Food Scientists and Technologists (AFSTI), Mysore, India, and the Punjab Academy of Sciences. She also serves as a reviewer for various international journals. She was awarded a postdoctoral fellowship from Korea University, Seoul, South Korea. Dr. Kaur's specialization is the characterization of bio-macromolecules from different botanical sources, their modifications and applications. In addition, Dr. Kaur has supervised three Ph.D. students.

# Contributors

**Ms. Amanjyoti**
Department of Food Science and
  Technology
Chaudhary Devi Lal University
Sirsa (Haryana), India

**Dr. Supriya Ambawat**
ICAR—All India Coordinated
  Research Project on Pearl Millet
Agriculture University
Jodhpur

**Dr. Vandana Chaudhary**
College of Dairy Science and
  Technology
Lala Lajpat Rai University of Veter-
inary and Animal Sciences
Hisar (Haryana), India

**Dr. Sanju Bala Dhull**
Department of Food Science and
  Technology
Chaudhary Devi Lal University
Sirsa (Haryana), India

**Professor Suresh Kumar Gahlawat**
Department of Biotechnology
Chaudhary Devi Lal University
Sirsa (Haryana), India

**Dr. Priyanka Kajla**
Deapartment of Food Technology
Guru Jambheshwar University of
  Science & Technology
Hisar (Haryana), India

**Dr. Maninder Kaur**
Department of Food Science and
  Technology
Guru Nanak Dev University
Amritsar (Punjab), India

**Ms Pinderpal Kaur**
Department of Food Science and
  Technology
Maharaja Ranjit Singh Punjab Tech-
  nical University
Bathinda (Punjab), India

**Ms Shamandeep Kaur**
Department of Food Science and
  Technology
Maharaja Ranjit Singh Punjab Tech-
  nical University
Bathinda (Punjab), India

**Dr. R.C. Meena**
ICAR—All India Coordinated
  Research Project on Pearl Millet
Agriculture University
Jodhpur

**Dr. Manju Nehra**
Department of Food Science and
  Technology
Chaudhary Devi Lal University
Sirsa (Haryana), India

**Dr. Kawaljit Singh Sandhu**
Department of Food Science and
  Technology
Maharaja Ranjit Singh Punjab Tech-
  nical University
Bathinda (Punjab), India

**Dr. Sneh Punia**
Department of Food Science and
  Technology
Chaudhary Devi Lal University
Sirsa (Haryana), India

**Dr. C. Tara Satyavathi**
ICAR—All India Coordinated
  Research Project on Pearl Millet
Agriculture University
Jodhpur

**Dr. Sukhvinder Singh Purewal**
Department of Food Science and
  Technology
Maharaja Ranjit Singh Punjab Tech-
  nical University
Bathinda (Punjab), India

**Dr. Loveleen Sharma**
Amity Institute of Food Technology
Amity University Noida
Uttar Pradesh, India

**Mr. Shobhit**
Department of Food Technology
Guru Jambheshwar University of
  Science & Technology
Hisar (Haryana) India

**Dr. Ajay Singh**
Department of Food Technology
Mata Gujri College
Fatehgarh Sahib (Punjab), India

**Dr. Subaran Singh**
Agriculture University
Jodhpur (Rajasthan) India

**Dr. Anil Kumar Siroha**
Department of Food Science and
  Technology
Chaudhary Devi Lal University
Sirsa (Haryana), India

**Dr. Suman**
Department of Foods & Nutrition
CCS Haryana Agricultural
  University
Hisar (Haryana) India

# Abbreviations

| | |
|---|---|
| **AFLPs** | amplified fragment length polymorphisms |
| **AM** | *Arbuscular mycorrhiza* |
| **BHA** | butylated hydroxyanisole |
| **BHT** | butylated hydroxytoluene |
| **BV** | breakdown viscosity |
| **Ca** | calcium |
| **CISP** | conserved intron spanning primer |
| **cP** | centi Pascal |
| **CRISPR/Cas** | clustered regularly interspaced short palindromic repeat/ CRISPR-associated protein) |
| **DArT** | diversity arrays technology |
| **DC** | degree of cross linking |
| **DS** | degree of substitution |
| **EPI** | epichlorohydrin |
| **EST** | expressed sequence tag |
| **FAO** | Food and Agricultural Organisation |
| **Fe** | iron |
| **FFA** | free fatty acids |
| ***ft** | feet |
| **FV** | final viscosity |
| **g/g** | gram per gram |
| **G'** | storage modulus |
| **G''** | loss modulus |
| **GBS** | genotyping-by-sequencing |
| **GC-TOFMS** | gas chromatography time of flight mass spectrometry |
| **GE** | genetic engineering |
| **GGT** | genomics guided transgenes |
| **GI** | glycemic index |
| **GMOs** | genetically modified organisms |
| **GWAS** | genome wide association studies |
| **h** | hour |
| **HCL** | hydrochloric acid |
| **HMT** | heat moisture treatment |
| **IAEA** | International Atomic Energy Agency |
| **ICRISAT** | International Crop Research Institute for Semi Arid Tropics |
| **K** | consistency index |
| **KJ** | kilo joule |
| **LGF** | large granule fraction |
| **MAB** | marker assisted breeding |
| **MAGIC** | multiparent advanced generation intercross |
| **MAS** | marker assisted selection |
| **mg/kg** | milligram/kilogram |

| | |
|---|---|
| **mha** | million hectares |
| **mPa.s** | milli Pascal second |
| **n** | flow behaviour index |
| **NAM** | nested association mapping |
| **NGS** | next-generation sequencing |
| **OPVs** | open pollinated varieties |
| **OSA** | octenyl succinic anhydride |
| **Pa—** | Pascal |
| **PGPR** | plant growth promoting rhizobacteria |
| **PGRs** | plant growth regulators |
| **pH** | potential of hydrogen |
| **PMiGAP** | inbred Germplasm Association Panel of Pearl millet |
| **POCl$_3$** | phosphoryl chloride |
| **Ppm** | parts per million |
| **PPO** | polyphenol oxidase |
| **PT** | pasting temperature |
| **PV** | peak viscosity |
| **q/ha** | quintals/hectare |
| **QTL** | quantitative trait loci |
| **RAD-seq** | restriction site-associated DNA sequencing |
| **RDA** | recommended dietary allowances |
| **RDS** | rapidly digestible starch |
| **RFLP** | restriction fragment length polymorphism |
| **RIIC** | Rural Industries Innovation Center |
| **RS** | resistant starch |
| **SAP** | steaming after pearling |
| **SBP** | steaming before pearling |
| **SCAR** | sequence characterized amplified region |
| **SDS** | slowly digestible starch |
| **SEM** | scanning electron microscopy |
| **SGF** | small granule fraction |
| **SNP** | single nucleotide polymorphism |
| **SP** | swelling power |
| **SSCP** | single stranded conformational polymorphism |
| **SSRs** | simple sequence repeats |
| **STMP** | sodium trimetaphosphate |
| **STPPv** | sodium tripolyphosphate |
| **STSs** | sequence tagged sites |
| **SV** | setback viscosity |
| **TALENs** | transcriptional activator-like effector nucleases |
| **tan** | δ damping factor |
| **Tc** | endset temperatures |
| **TILLING** | targeted induced local lesion in genome |
| **To** | onset temperatures |
| **Tp** | peak temperatures |
| **TV** | trough viscosity |

| | |
|---|---|
| **w/v** | weight by volume |
| **w/w** | weight by weight |
| **ZFNs** | zinc finger nucleases |
| **Zn** | zinc |
| **\*β** | beta |
| **ΔHgel** | enthalpy of gelatinization |
| **η\*** | complex viscosity |
| **\*θ** | theta |
| **\*μm** | micro meter |
| **σo** | yield stress |

# Preface

Pearl millet is gaining importance as a climate-resilient and health promoting nutritious crop. It is mainly used for animal and poultry feed and, therefore, has low cost. By exploring the properties of pearl millet, its utilization can be extended to more food applications. This book explores knowledge about pearl millet production, grain structure, chemistry and nutritional aspects, primary processing technologies, and product formulations.

Chapter 1 deals with pearl millet classification, history, nutritional aspects, and health benefits. Pearl millet has comparable nutritional content to other cereals grains, so it can be substituted for cereals such as wheat and corn. Pearl millet is prone to rancidity, therefore different methods to increase the shelf life of pearl millet flour are discussed in Chapter 2. Chapter 3 discusses the phytochemical and antioxidant properties of pearl millet flour, as well as the different components which are responsible for the antioxidant properties of pearl millet.

Various processing methods to increase the nutritional properties of pearl millet are discussed in Chapter 4. These include decortication, blanching, acid treatment, germination, fermentation, microwave cooking, extrusion cooking, roasting and toasting. Chapter 5 deals with starch structure, properties and applications. In this chapter, starch isolation methods, starch structure, pasting, rheological, morphological, digestibility properties and the application of pearl millet starch are elaborated. Pearl millet starch has comparable properties to other cereal starches, such as corn starch, so can be used as a substitute for such starches.

During processing, the texture and appearance of the product is altered, so, to overcome these undesirable changes, starch needs to be modified. Different methods of starch modification, such as physical, chemical and enzymatic, are explored in Chapter 6. Chapter 7 deals with biotechnological applications for improvement of the pearl millet crop. In this chapter, conventional approaches for genetic improvement, biotechnological approaches, tissue culture, and transgenic/transformation approaches, recombinant DNA technology, molecular markers, QTL mapping, and genetic modifications/engineering are discussed. Chapter 8 explores the biofortification and medicinal value of pearl millet. Pearl millet cultivars containing higher amounts of iron and zinc are produced to overcome malnutrition. Various components are present in pearl millet, such as iron, zinc, magnesium, phosphorous, calcium and antioxidants which are responsible for its medicinal properties.

Chapter 9 focuses on various products formulated from pearl millet. In this chapter, various traditional, fermented and new products from pearl millet are discussed.

This book can be useful for the students, academicians, researchers and other interested professionals working in starch, antioxidants, and new product formulations. There are many books available on pearl millet, but this book is designed in such a way that it deals with important aspects related to pearl millet. The editors would appreciate receiving any comments/information that might be helpful for the future course of action with regard to this product.

# 1 Pearl Millet
## A Drought Arrested Crop

*Sneh Punia, Anil Kumar Siroha, Kawaljit Singh Sandhu, Suresh Kumar Gahlawat and Maninder Kaur*

## CONTENTS

## 1.1 INTRODUCTION

**Taxonomy**
***Pennisetumglaucum*** (Pearl Millet)
**Kingdom**: Plantae—plants
**Subkingdom**: Tracheobionta—vascular plants
**Superdivision**: Spermatophyta—seed plants
**Division**: Magnoliophyta—flowering plants
**Class**: Liliopsida—monocotyledons
**Subclass**: Commelinidae
**Order**: Cyperales
**Family**: Poaceae—grass family
**Genus**: Pennisetum—fountaingrass
**Species**: Pennisetum glaucum—pearl millet

In the 21st century, climate changes, water scarcity, increasing world population, rising food prices, and other socioeconomic impacts are expected to generate a great threat to agriculture and food security worldwide, especially for the poorest people who live in arid and sub-arid regions. These impacts present a challenge to scientists and nutritionists to investigate the possibilities of producing, processing and utilizing other potential food sources to end hunger and poverty (Saleh et al., 2013). Cereals are annual common grass members of the family Poaceae, also known as Gramineae, and the major ones are wheat, barley, maize, oats, rice, rye, sorghum, and millet (Wrigley, 2016). The global production of cereals is about 2,723,878,753 tonnes whereas it was 257,499,500 tonnes in India (FAO, 2017). They provide significant amounts of energy, protein, and micronutrients in both animal and human diets and are considered to be one of the most important sources of dietary proteins, carbohydrates, vitamins, minerals and fiber for people all over the world (Dordevic et al., 2010). Millets are one of the cereals, besides wheat, rice, and maize, that are major food sources for millions of people, especially those who live in hot, dry areas of the world. They are grown mostly in marginal areas under agricultural conditions in which other cereals fail to give substantial yields (Adekunle, 2012). Millets are classified (along with maize and sorghum) in the grass sub-family Panicoideae (Yang et al., 2012). They are small-seeded cereals that are cultivated as subsistence crops, mainly in semi-arid and tropical regions in Asia and Africa where they are widely grown because they are one of the most important drought-resistant crops. Millet constitutes a major source of carbohydrates and proteins for people living in these areas. In addition, because of its important contribution to national food security and potential health benefits, millet grain is now receiving increasing interest from food scientists, technologists, and nutritionists (Saleh et al., 2013). The total world production of millet grains was 28,357,451 tonnes and the top producer was India, with an annual production of 10,280,000 tonnes (FAOSTAT, 2016). They have been reported to be nutritionally superior when compared to many other cereals (Parameswaran and Sadasivam, 1994; Saleh et al., 2013) and are good sources of proteins, carbohydrates, fiber, essential amino acids, phytochemicals, and micronutrients (Singh et al., 2012). Several species of millets exist: these include pearl (*Pennisetum glaucum*), browntop (*Brachiaria ramose*), barnyard (*Echinochloa crusgalli*), finger (*Eleusine coaracana*), proso (*Panicum miliaceum*), kodo (*Paspalum scrobiculatum*), little (*Panicum sumatrense*) and foxtail (*Setaria italica*). Among these millet species, pearl millet is the most widely cultivated crop (Saleh et al., 2013). Pearl millet (*Pennisetum glaucum* (L.) is also classified as *P. typhoides*, *P. americanum* or *P. spicatum*) and locally known as *mahangu*, *bajra* and *dukhon*. Pearl millet is a cereal grass cultivated almost exclusively on a subsistence basis by farmers in semi-arid parts of Africa and Asia (ICRISAT and FAO, 1996; Jain and Bal, 1997; Taylor, 2004).

## 1.2  HISTORY

Millet is one of the oldest cultivated foods known to humans (Oelke et al., 1990). Since prehistoric times, pearl millet has been grown prominently in Africa and the Indian subcontinent. It is believed that pearl millet originated in Africa and was later introduced to India. The earliest archaeological evidence shows that millet was cultivated in India around 2,000 BC. Millet is an important food staple in Africa, where it is used to make a traditional flatbread known as *injera*. The ground millet seeds are used for making the Indian flatbread called *roti*. In the Middle Ages, millet was a staple grain in Europe, especially in countries in Eastern Europe. It was introduced into the United States in the 19th century. Millet is popular as birdseed and livestock fodder in Western Europe and North America, but it has recently gained popularity as a delicious and nutritious grain due to its nutritional benefits and gluten-free status. Currently, India is the leading commercial producer of pearl millet, followed by China and Nigeria.

## 1.3  PEARL MILLET PLANT DESCRIPTION

Pearl millet is a very robust grass which tills widely and grows in tufts. It has slender stems which are divided into distinct nodes. The leaves of the plant are linear, or lance-like, possess small teeth, and can grow up to 1 m (3.3 ft) in length. The inflorescence of the plant is a spike-like panicle, made up of many smaller spikelets where the grain is produced. Pearl millet can reach 0.5 to 4 m (1.6–13.1 ft) in height, depending on the cultivar, and is an annual plant, harvested after one growing season (Figure 1.1).

**FIGURE 1.1**  Pearl millet plant

## 1.4 PEARL MILLET GRAIN DESCRIPTION

Grains thresh free of the hull and vary from pearly white to yellow in color (Taylor, 2004), shaped like a liquid drop (Jain and Bal, 1997), light in weight (3–15 mg) but have a proportionally larger germ (17.4%) than all other cereal grains, except maize (Taylor, 2004). The grains can be up to 2 mm in length, with weight ranging between 3 mg and 15 mg. The grains are packed closely together and have a density of 1.6 g/cm$^3$ (Serna-Saldivar, 1995; Jain and Bal, 1997), which is significantly higher than that of wheat (1.39 g/cm$^3$), maize (1.39 g/cm$^3$), rice (1.24 g/cm$^3$) and sorghum (1.24 g/cm$^3$) grains (Serna-Saldivar, 1995). The pearl millet grain comprises pericarp (8%), germ (17%) and endosperm (75%) (Dendy, 1995). A thin, waxy cutin layer covers the surface of the pericarp, which helps to decrease the effect of weathering. Beneath the pericarp, a thin layer of seed coat and a single aleurone layer are present. Bran is the hard outer layer of cereals which consists of a combination of aleurone, pericarp, and the part of the germ that is rich in dietary fiber, essential fatty acids, protein (17%), oil (32%), ash (10.4%) and starch, vitamins, and dietary minerals. Pearl millet is rich in micro-nutrients, such as iron (Fe) and zinc (Zn), and has an appreciable level of antioxidant compounds (Berwal et al., 2016).

## 1.5 NUTRITIONAL DESCRIPTION

Pearl millet is the most widely grown millet type followed by foxtail, proso, and finger millets. It is significantly rich in resistant starch, soluble and insoluble dietary fibers, minerals, and antioxidants (Ragaee et al., 2006). It contains dry matter (92.5%), ash (2.1%), crude fiber (2.8%), crude fat (7.8%), crude protein (13.6%), and starch (63.2%) (Ali et al., 2003). Sade (2009) reported 14.0, 5.7, 2.1, 2.0 and 76.3 g/100 g of crude protein, fat, ash, crude fiber and carbohydrate content, respectively, in pearl millet. The proteins have a better amino acid profile than those of many other cereals, such as corn and sorghum (FAO, 1970). Han et al. (1978) reported that among non-essential amino acids, glutamic acid, alanine, aspartic acid and proline, and essential amino acids, luecine and valine, are present in large quantities in pearl millet. These results agree with those of Sawhney and Naik (1969). Further, Sawaya et al. (1984) reported that glutamic acid, leucine, aspartic acid and alanine are the major amino acids in pearl millet protein. The protein quality of pearl millet is low in the levels of lysine and tryptophan (FAO, 1995). Lysine is by far the most limiting amino acid in diets containing pearl millet as the only source of protein (Adrian and Sayerse, 1957; Goswami et al., 1969; Jones et al., 1970; Swaminathan et al., 1971; Burton et al., 1972; Awadalla and Slump, 1974; Badi et al., 1976). Compared with sorghum (Badiet al.,1976), corn, and wheat (Mangay et al., 1957), pearl millet had higher values for histidine, threonine, valine and phenylalanine than sorghum, corn and wheat, higher values of lysine and isoleucine than sorghum and wheat, higher values of methionine than sorghum, higher values of leucine than corn and wheat, and lower values of methionine than corn and wheat, lower values of isoleucine than corn, and lower values of leucine than sorghum (Han, 1978).

**TABLE 1.1**

**Nutritional composition of pearl millet seed**

| Nutrients | Composition |
|---|---|
| Carbohydrate | 71.6 (Taylor, 2004) |
| Protein | 13 |
| Fat | 6.3 |
| Fibre | |
| Ash | 17 |
| **Mineral (mg/100 g)** | |
| Iron | 1.9–5.5 |
| Zinc | 2–.2 |
| Calcium | 7.2–.4 |
| Phosphorus | 326–73 |
| Magnesium | 137 |
| **Vitamins (mg/100 g)** | |
| Vitamin A | 24 |
| Thiamine | 0.3 |
| Riboflavin | 0.2 |
| Niacin | 2.9 |
| Vitamin E | 1.9 |
| **Essential amino acid (g/100 g)** | |
| Phe | 5.5 |
| Ile | 3.1 |
| Leu | 9.6 |
| Lys | 3 |
| Met | 2.3 |
| Thr | 3.6 |
| Val | 3.9 |
| **Non-essential amino acid (g/100 g)** | |
| Asp | 8.8 |
| Glu | 19.6 |
| Ala | 8.2 |
| Arg | 4.8 |
| Gly | 3.3 |
| Pro | 6.4 |
| Ser | 5.2 |
| Tyr | 3.8 |

Source: (Serna-Saldivar, 1995; Taylor, 2004; Lestienne et al., 2007; Bashir et al., 2014; Minnis-Ndimba et al., 2015)

The storability of pearl millet is poor due to its high fat content (Lai &Varriano-Marston, 1980b). About 88% of the lipid fraction is found in the germ (Abdelrahman et al., 1983), whereas pericarp and endosperm contain 6% each. Pearl millet has a high content of triglycerides and polyunsaturated fatty acids that constitute the triglycerides and negatively affect the shelf life of untreated pearl millet flour (Chaudhary and Kapoor, 1984). However, while the grain is still whole, it can be stored for 6–8 months in dry regions, but once crushed, its grits and flour have poor storability (Lai &Varriano-Marston, 1980b) and turn bitter and rancid within a few days of being stored due to lipolysis and subsequent oxidation of unsaturated fatty acids (Lai and Varriano-Marston, 1980a). This is caused by an active lipase enzyme that is responsible for the breakdown of glycerides and consequent increase of free fatty acids (Pruthi, 1981; Arora et al., 2002). These chemical changes are usually manifested as off-flavors during storage, especially under conditions of moderately high moisture and oxygen exposure (Nantanga et al., 2008).

According to Osagie and Kates (1984), the fat in pearl millet grain consists of 85% neutral (non-polar) lipids, 12% phospholipids and 3% glycolipids. Neutral lipids comprise about 85% triglycerides and small amounts of mono- and diglycerides, sterols and free fatty acids. Sterols comprise camesterol and stigmasterol. Lysophosphatidylcholine, phosphatidylcholine and lysophosphatidylethanolamine are the major phospholipids. These researchers also found that the major glycolipids were esterified sterol glycoside, sterol glucoside and mono- and digalactosyldiacylglycerol. Pearl millet triglycerides contain about 74% unsaturated fatty acids, mainly oleic (C18:1), linoleic (C18:2) and linolenic (C18:3) acids (Rooney, 1978; Lai and Varriano-Marston, 1980a; Kapoor and Kapoor, 1990). The remaining fraction is made up of saturated fatty acid residues (palmitic (C16:0) and stearic (C18:0)). Omega 3, linolenic acid (C18:3n-3) (LNA), comprises 4% of the total fatty acids in this oil (Burton et al., 1972; Rooney, 1978).

## TABLE 1.2
## Composition of fat of pearl millet

| Neutral lipids (85%) | Phospholipids (12%) | Glycolipids (3%) |
|---|---|---|
| Steryl ester | Lysophosphatidylcholine (LPC) | Acyl-monogalactosyldiacylglycerol |
| trace - Triacylglycerols | Lysophosphatidylethanolamine (LPE) | Esterified sterolglycoside |
| Mono-and diacylglycerols | Phosphatidylcholine (PC) | Monogalactosyldiacylglycerol |
| trace Free fatty acids | Phosphatidylethanolamine (PE) | Unidentified |
| Free sterols | Phosphatidylglycerol (PG) | Sterol glycoside |
| | Phosphatidic acid (PA) | Cerebroside I |
| | Biphosphatidic acid (X 4) | Cerebroside II |
| | | Monogalactosylmonoacylglycerol |
| | | Digalactosyldiacylglycerol |
| | | Digalactosylmonoacylglycerol |

Source: Osagie and Kates (1984)

The ash content of whole pearl millet grain ranges between 1.6% and 3.6% (Serna-Saldivar, 1995). In terms of actual minerals, pearl millet grain, like other cereal grains, is an adequate source of dietary minerals such as magnesium, iron, zinc and copper (Nantanga, 2007). Pearl millet is a good source of minerals, containing appreciable amounts of calcium, phosphorus, magnesium, and iron (Burton et al., 1972).

Like other cereal grains, pearl millet grain is an important source of thiamin, niacin, and riboflavin (Taylor, 2004). Riboflavin has, however, been implicated in lipid deterioration in the presence of light (Hamilton, 1999). Thus, it may also be a potential enhancer of the deterioration of pearl millet triglycerides. Because of its high oil content, pearl millet is also a good source of lipid-soluble vitamin E. Its content in pearl millet is about 2 mg/100 g (Taylor, 2004). Millet contains tannins (0.61), phytates (0.48%), polyphenols, trypsin inhibitors, and dietary fiber which are considered as "anti-nutrients" because of their metal chelating and enzyme inhibitor activities, which are termed nutraceuticals (Thompson, 1993). Biochemical composition of pearl millet indicates that it is a good source of energy, protein, vitamins, and minerals (Osman, 2009). However, bioavailability of the nutrients is restricted due to the presence of anti-nutritional factors such as phytic acid, tannins, goitrogens, oxalic acid and trypsin inhibitors. These compounds interfere with mineral bioavailability, carbohydrates and protein digestibility through inhibition of proteolytic and amylolytic enzymes (Jalgaonkar et al., 2016). Tannins are polyphenolic compounds which bind to proteins, carbohydrates, and minerals, thereby reducing the digestibility of these nutrients (Dykes and Rooney, 2006).

The phytic acid is present in the germ, whereas polyphenols are in the peripheral areas of the pearl millet grain (Simwemba et al., 1984). Phytic acid has a strong ability to chelate multivalent metal ions, especially zinc, calcium, iron, and as with protein residue. The binding can result in insoluble salts with poor bioavailability of minerals (Coulibaly et al., 2011). Phytate renders minerals, especially divalent cations, unavailable and inhibits proteolytic and amylolytic enzymes (Sutardi-Buckle, 1985). Polyphenol content of pearl millet is fairly high, which affects the mineral bioavailability and protein and carbohydrate digestibility of food grains (Thompson and Yoon, 1984; Pawar and Parlikar, 1990). Polyphenols also impart a gray colour to flour, besides limiting its nutrient utilization (Pawar and Parlikar, 1990). Proteinaceous inhibitors of alpha amylases are widely distributed in cereals, legumes, and other plants, including pearl millet (Sharma and Pattabiraman, 1982; Udupa et al., 1989; Feng et al., 1991; Henry et al., 1992) and these reduce starch utilization.

Although luteolin, a flavone present in millets, has antioxidant properties, other studies have shown evidence that the presence of C-glycosyl flavones in foods may promote the inhibition of thyroid peroxidase (TPO), an enzyme produced in the thyroid gland responsible for the production of thyroid hormones (Gaitan et al., 1989; Mezzomo and Nadal, 2016).

Starch is the most abundant reserve carbohydrate of plants. It is a macro-constituent of many foods and its properties and interactions with other constituents, particularly water and lipids, are of interest to the food industry and for human nutrition (Copeland et al., 2009). Starch is the main source of energy in the diet of humans. Therefore, a detailed investigation is necessary for better understanding of its biochemical and functional characteristics (Kaur et al., 2004). Starch is made up of two fractions: amylose, which is made up of essentially $\alpha$-(1/4) D-glucopyranosyl units, and amylopectin, which is made up of a large number of short chains linked together at their reducing end by a $\alpha$-(1/6) linkage (Biliaderis, 1998). Identification of native starch sources is required for desired functionality and unique properties (Duxbury, 1989). Differences among rheological properties of starch water solutions depend on amylose and amylopectin content, the presence of functional groups (i.e. phosphates), and also granularity (Berski et al., 2011).

Starch continues to play a special role in the food industry. In addition to its nutritional value, starch is employed in many applications as a thickener, binder, and gelling agent. Non-food applications of starch include its use in the field of pharmaceuticals, textiles, alcohol-based fuels, and adhesives. New uses of starch include low-calorie substitutes, biodegradable packaging materials, thin film, and thermoplastic materials with improved thermal and mechanical properties (Biliaderis, 1998). Most regular starch pastes do not possess the application properties desirable in food processing, therefore there is a need to screen starches from new carbohydrate sources for novel properties (Nwokocha and Williams, 2011).

The applications of native starch are limited due to its low solubility in cold water, limited emulsification capability, easy retrogradation, poor storage stability under refrigerated and processing conditions, variable shearing force, and pH values (Singh et al., 2007). Therefore, starch used in the food industry is often modified to overcome undesirable changes in product texture and appearance caused by retrogradation or breakdown of starch during processing and storage (Van Hung and Morita, 2005). Native starches contain free hydroxyl groups in the 2, 3 and 6 carbons of the glucose molecule, making them highly reactive. This allows them to be modified by different chemical treatments and, thus, regulate their properties (Bao et al., 2003). The chemical modification of starch can be achieved by a variety of different chemical reactions, such as acid hydrolysis, oxidation, etherification, esterification, and cross-linking (Jayakody and Hoover, 2002).

## 1.6  HEALTH BENEFITS

Pearl millet serves as a major staple food for many populations around the globe; however, it is still considered poor man's food and tends not to be included in the diets of the elite (Nambiar et al., 2011). Pearl millet is a gluten-free grain and is the only grain that retains its alkaline properties after being cooked, which is ideal for people with wheat allergies. The concept of glycemic index (GI) emerged as a physiological basis for ranking carbohydrate foods according to the blood glucose response they produce on ingestion, and was introduced by Jenkins et al. (1978). Mani et al. (1993) have reported that pearl millet has the lowest GI (55) as compared to some other cereals. Foods with a low glycemic index are useful to manage maturity onset diabetes, by improving metabolic control of blood pressure and plasma low density lipoprotein cholesterol levels due to less pronounced insulin response (Asp, 1996). Pearl millet contains high levels of antioxidants, namely the phenolic compounds, and may have anticancer properties. Phenolic compounds, especially flavanoids, have been found to inhibit tumor development (Huang and Ferraro, 1992). A large number of clinical studies have recognized the tremendous potential of omega-3 fatty acids. The presence of omega-3 fatty acids in pearl millet highlights its potential in the prevention and treatment of cardiovascular diseases, diabetes, arthritis, and certain types of cancer (Nambiar et al., 2011).

## REFERENCES

Abdelrahman, A., Hoseney, R. C. and Varriano-Marston, E. 1983. Milling process to produce lowfat grits from pearl millet. *Cereal Chemistry* 60: 189–191.

Adekunle, A. A. 2012. Agricultural innovation in sub-Saharan Africa: Experiences from multiple stake holder approaches. Forum for Agricultural Research in Africa, Ghana. ISBN 978-9988-8373-2-4.

Adrian, J. and Sayerse, C. 1957. Composition of Senegal millets and sorghums. *British Journal of Nutrition* 11: 99–105.

Aggarwal, A. J. 1992. Processing of pearl millet for its more effective utilisation. Ph.D. Thesis CCS Haryana Agricultural University, Hisar, India.

Ali, M. A., El Tinay, A. H. and Abdalla, A. H. 2003. Effect of fermentation on the in vitro protein digestibility of pearl millet. *Food Chemistry* 80(1): 51–54.

Arora, P., Sehgal, S. and Kawatra, A. 2002. The role of dry heat treatment in improving the shelf-life of pearl millet flour. *Nutrition and Health* 16: 331–336.

Asp, N. G. 1996. Dietary carbohydrate: Classification by chemistry and physiology. *Journal of Food Chemistry* 7: 9–14.

Awadalla, M. and Slump, P. 1974. Native Egyptian millet as supplement of wheat flour in bread: I. Nutritional studies. *Nutrition Reports International* 9: 59.

Badi, S. M., Hoseney, R. C. and Casady, A. J. 1976. Pearl millet. I. Characterization by SEM, amino acid analysis, lipid composition, and prolamine solubility. *Cereal Chemistry* 53(4): 478.

Bao, J., Xing, J., Phillips, D. L. and Corke, H. 2003. Physical properties of octenyl succinic anhydride modified rice, wheat, and potato starches. *Journal of Agricultural and Food Chemistry* 51: 2283–2287.

Bashir, E. M., Ali, A. M., Ali, A. M., Melchinger, A. E., Parzies, H. K. and Haussmann, B. I. 2014. Characterization of Sudanese pearl millet germplasm for agro-morphological traits and grain nutritional values. *Plant Genetic Resources* 12(1): 35–47.

Berski, W., Ptaszek, A., Ptaszek, P., Ziobro, R., Kowalski, G., Grzesik, M. and Achremowicz, B. 2011. Pasting and rheological properties of oat starch and its derivatives. *Carbohydrate Polymers* 83: 665–671.

Berwal, M. K., Chugh, L. K., Goyal, P. and Kumar, R. 2016. Total antioxidant potential of pearl millet genotypes: Inbreds and designated b-lines. *Indian Society of Agricultural Biochemists* 29: 201–204.

Biliaderis, C. G. 1998. Structures and phase transitions of starch polymers. In: R. H. Walter (Ed.), *Polysaccharide association structures in food*, New York: Marcel-Dekker, pp. 57–168.

Burton, G. W., Wallace, A. T. and Rachie, K. O. 1972. Chemical composition and nutritive value of pearl millet (*Pennisetumtyphoides*) grain. *Crop Science* 12: 187–188.

Chandrasekara, A., Naczk, M. and Shahidi, F. 2012. Effect of processing on the antioxidant activity of millet grains. *Food Chemistry* 133(1): 1–9.

Chaudhary, P. and Kapoor, A. C. 1984. Changes in the nutritional value of pearl millet flour during storage. *Journal of the Science of Food and Agriculture* 35: 1219–1224.

Copeland, L., Blazek, J., Salman, H. and Tang, M. C. 2009. Form and functionality of starch. *Food Hydrocolloids* 23: 1527–1534.

Coulibaly, A., Kouakou, B. and Chen, J. 2011. Phytic acid in cereal grains: Structure, healthy or harmful ways to reduce phytic acid in cereals grains and their effects on nutritional quality. *American Journal of Plant Nutrition and Fertilization Technology* 1(1): 1–22.

Dendy, D. A. V. 1995. *Sorghum and Millets: Chemistry and technology*, St. Paul, MN: American Association of Cereal Chemists, pp. 223–281.

Dordevic, T. M., Siler-Marinkovic, S. S. and Dimitrijevic-Brankovic, S. I. 2010. Effect of fermentation on antioxidant properties of some cereals and pseudo cereals. *Food Chemistry* 119: 957–963.

Duxbury, D. D. 1989. Modified starch functionalities—no chemicals or enzymes. *Food Processing* 50: 35–37.

Dykes, L. and Rooney, L. W. 2006. Sorghum and millet phenols and antioxidants. *Journal of Cereal Science* 44(3): 236–251.

FAO. 1970. Amino acid content of foods and biological data on proteins. Nutritional Studies No, 24, Rome: Food and Agriculture Organization.

FAO. 1995. Sorghum and millets in human nutrition. (FAO Food and Nutrition Series, No. 27), Rome, Italy: Food and Agriculture Organization of the United Nations.

FAOSTAT. 2016. Food and agriculture organisation of the United Nations. FAOSTAT Database. http://faostat3.fao.org/download/Q/QI/E. Accessed on 07/03/2016.

FAOSTAT 2017. Food and Agriculture Organisation of the United Nations. FAOSTAT Database. http://faostat.fao.org/beta/en/#data/QC. Accessed on 30/10/2017.

Feng, G. H., Chen, M., Kramer, K. J. and Reeck, G. R. 1991. Alpha-amylase inhibitors from rice: Fractionation and selectivity toward insect, mammalian and bacterial alpha-amylase. *Cereal chemistry* 68(5): 516–521.

Gaitan, E., Lindsay, R. H., Reichert, R. D., Ingbar, S. H., Cooksey, R. C., Legan, J. and Kubota, K. 1989. Antithyroid and goitrogenic effects of millet: Role of C-glycosylflavones. *Journal of Clinical Endocrinology & Metabolism* 68(4): 707–714.

Goswami, A. K., Sharma, K. P. and Sehgal, K. L. 1969. Nutritive value of proteins of pearl millet of high-yielding varieties and hybrids. *British Journal of Nutrition* 23: 913.

Hamilton, R. J. 1999. The chemistry of rancidity in foods. In: J. C. Allen and R. J. Hamilton (Eds.), *Rancidity in foods*. Gaithersburg, MD: Aspen Publishers, pp. 1–21.

Han, R. T. P. 1978. Genotypic variability in protein content and amino acid composition of pearl millet (*PennisetumTyphoideum*) (Doctoral dissertation, Texas Tech University).

Henry, R. J., Battershell, V. G., Brennan, P. S. and Oono, K. 1992. Control of wheat alpha-amylase by using inhibitors from cereals. *Journal of the Science of Food and Agriculture* 58: 281–284.

Huang, M. T. and Ferraro, T. 1992. Phenolics compounds in food and cancer prevention. *Phenolic Compounds in Food and Their Effects on Health II, ACS Symposium Series* 507: 8–34.

ICRISAT and FAO. 1996. *The world sorghum and millet economies.* International Crops Research Institute for the Semi-Arid Tropics, Patancheru, India, Food and Agriculture Organisation of the United Nations, Rome, pp. 31–53.

Jain, R. K. and Bal, S. 1997. Properties of pearl millet. *Journal of Agricultural Engineering Research* 66: 85–91.

Jalgaonkar, K., Jha, S. K. and Sharma, D. K. 2016. Effect of thermal treatments on the storage life of pearl millet (*Pennisetumglaucum*) flour. *Indian Journal of Agricultural Sciences* 86(6): 762–767.

Jayakody, L. and Hoover, R. 2002. The effect of linterization on cereal starch granules. *Food Research International* 35: 665–680.

Jenkins, D. J. A., Wolever, T. M. S. and Leeds, A. R. 1978. Dietary fibres, fibre analogues and glucose tolerance: Importance of viscosity. *British Medical Journal* 1: 392–394.

Jones, R. W., Beckwith, A. C., Khoo, U. and Inglett, G. E. 1970. Protein composition of proso millet. *Journal of Agriculture and Food Chemistry* 18(1): 37–39.

Kapoor, R. and Kapoor, A. C. 1990. Effect of different treatments on keeping quality of pearl millet flour. *Food Chemistry* 35: 277–286.

Kaur, M., Singh, N., Sandhu, K. S. and Guraya, H. S. 2004. Physicochemical, morphological, thermal and rheological properties of starches separated from kernels of some Indian mango cultivars (*Mangifera indica L.*). *Food Chemistry* 85: 131–140.

Lai, C. C. and Varriano-Marston, E. 1980a. Lipid content and fatty acid composition of free and bound lipids in pearl millet. *Cereal Chemisty* 57: 271–274.

Lai, C. C. and Varriano-Marston, E. 1980b. Changes in the pearl millet during storage. *Cereal Chemistry* 57: 275.

Lestienne, I., Buisson, M., Lullien-Pellerin, V., Picq, C., and Treche, S. 2007. Losses of nutrients and anti-nutritional factors during abrasive decortication of two pearl millet cultivars (*Pennisetumglaucum*). *Food Chemistry* 100(4): 1316–1323.

Mangay, A. S., Pearson, W. N., and Darby, W. J. 1957. Millet (Setariaitalica): Its amino acid and niacin content and supplementary nutritive value for corn (maize). *The Journal of Nutrition* 62(3): 377–393.

Mani, U. V., Prabhu, B. M., Damle, S. S., and Mani, I. 1993. Glycemic Index of some commonly consumed foods in Western India, Asia Pacific. *Journal of Clinical Nutrition* 2: 111–114.

Mezzomo, T. R. and Nadal, J. 2016. Efeito dos nutrientese substancias alimentares na funçao tireoidianae no hipotireoidismo. *DEMETRA: Alimentaçao, Nutriçao & Saúde* 11(2): 427–443.

Minnis-Ndimba, R., Kruger, J., Taylor, J. R., Mtshali, C., and Pineda-Vargas, C. A. 2015. Micro-PIXE mapping of mineral distribution in mature grain of two pearl millet cultivars. *Nuclear Instruments and Methods in Physics Research Section B: Beam Interactions with Materials and Atoms* 15(363): 177–182.

Nambiar, V. S., Dhaduk, J. J., Sareen, N., Shahu, T., and Desai, R. 2011. Potential functional implications of pearl millet (Pennisetumglaucum) in health and disease. *Journal of Applied Pharmaceutical Science* 1(10): 62.

Nantanga, K. K. M. 2007. Lipid stabilisation and partial pre-cooking of pearl millet by thermal treatments (Doctoral dissertation, University of Pretoria).

Nantanga, K. K. M., Seetharaman K., Kock, H. L., and Taylor, J. R. N. 2008. Thermal treatments to partially pre-cook and improve the shelf life of whole pearl millet flour. *Journal of the Science of Food and Agriculture* 88: 1892–1899.

Nwokocha, L. M. and Williams, P. A. 2011. Structure and properties of Treculia africana (*Decne*) seed starch. *Carbohydrate Polymers* 84: 395–401.

Oelke, E. A., Oplinger, E. S., Bahri, H., Durgan, B. R., Putnam, D. H., Doll, J. D., and Kelling, K. A. 1990. Rye. In: *Alternative Field Crops Manual*, University of Wisconsin: Ext. Serv., Madison, and University of Minesota, St. Paul.

Osagie, A. U. and Kates, M. 1984. Lipid composition of millet (*Pennisetumamericanum*) seeds. *Lipids* 19: 958–965.

Osman, M. A. 2009. Effect of germination on the nutritional quality of pearl millet. *Journal of Agricultural Science and Technology* 3(8): 1–6.

Parameswaran, K. P. and Sadasivam, S. 1994. Changes in the carbohydrates and nitrogenous components during germination of proso millet, (*Panicummiliaceum*). *Plant Foods for Human Nutrition* 45: 97–102.

Pawar, V. D. and Parlikar, G. S. 1990. Reducing the polyphenols and phytate and improving the protein quality of pearl millet by dehulling and soaking. *Journal of Food Science and Technology* 127: 140–143.

Pruthi, T. D. 1981. Free fatty acid changes during storage of bajra (*Penmselumtyphoideum*) flour. *Journal of Food Science and Technology* 18(6): 257–258.

Ragaee, S., Abdel-Aal, E. S. M., and Noaman, M. 2006. Antioxidant activity and nutrient composition of selected cereals for food use. *Food Chemistry* 98(1): 32–38.

Rooney, L. W. 1978. Sorghum and pearl millet lipids. *Cereal Chemistry* 55: 584–590.

Sade, F. O. 2009. Proximate, antinutritional factors and functional properties of processed pearl millet (Pennisetumglaucum). *Journal of Food Technology* 7(3): 92–97.

Saleh, A. S., Zhang, Q., Chen, J., and Shen, Q. 2013. Millet grains: Nutritional quality, processing, and potential health benefits. *Comprehensive Reviews in Food Science Food Safety* 12: 281–295.

Sarwar, M. H., Sarwar, M. F., Sarwar, M., Qadri, N. A., and Moghal, S. 2013. The importance of cereals (*Poaceae: Gramineae*) nutrition in human health: A review. *Journal of Cereals and Oilseeds* 4(3): 32–35.

Sawaya, W. N., Khalil, J. K., and Safi, W. J. 1984. Nutritional quality of pearl millet flour and bread. *Plant Foods for Human Nutrition* 34(2): 117–125.

Sawhney, S. K. and Naik, M. S. 1969. Amino acid composition of protein fractions of pearl millet and the effect of nitrogen fertilization on its proteins. *Indian Journal of Genetics and Plant Breeding* 29: 395–406.

Serna-Saldivar, S. 1995. Structure and chemistry of sorghum and millets. *Sorghum and Millets: Chemistry and Technology* 1995: 69–124.

Sharma, K. K. and Pattabiraman, T. N. 1982. Natural plant enzyme inhibitors. Purification and properties of an amylase-inhibitor from yam (*Diascoreaalata*). *Journal of the Science of Food and Agriculture* 33: 255–262.

Simwemba, C. G., Hoseney, R. C., Varrino-Marston, E., and Zeleznak, K. 1984. Certain B vitamin and phytic acid contents of pearl millet (*Pennisetumamericanum* L Leeke). *Journal of Agriculture and Food Chemistry* 32: 31–34.

Singh, J., Kaur, L., and McCarthy, O. J. 2007. Factors influencing the physico-chemical, morphological, thermal and rheological properties of some chemically modified starches for food applications – A review. *Food Hydrocolloids* 21: 1–22.

Singh, K., Mishra, A., and Mishra, H. 2012. Fuzzy analysis of sensory attributes of bread prepared from millet-based composite flours. *LWT-Food Science and Technology* 48: 276–282.

Sutardi-Buckle, K. A. 1985. Reduction in phytic acid levels in soybeans during tempeh production, storage and frying. *Journal of Food Science* 50: 260–261.

Swaminathan, M. S., Naik, M. S., Kaul, A. K., and Austin, A. 1971. Choice of strategy for the genetic upgrading of protein properties in cereals, millets and pulses. *Indian Journal of Agricultural Sciences* 41: 393.

Taylor, J. R. N. 2004. Millet: Pearl. In: C. Wrigley, H. Corke and C. E. Walker (Eds.), *Encyclopedia of grain science*, Vol. 2, London: Elsevier, pp. 253–261.

Thompson, L. U. 1993. Potential health benefits and problems associated with antinutrients with foods. *Food Research International* 26: 131–149.

Thompson, L. U. and Yoon, J. H. 1984. Starch digestibility as affected by polyphenol and phytic acid. *Journal of Food Science* 49: 1228–1229.

Udupa, S. L., Prabhakar, A. R., and Tandon, S. 1989. Alpha-amylase inhibitors in food stuffs. *Food Chemistry* 34: 95–101.

Van Hung, P. and Morita, N. 2005. Effect of granule sizes on physicochemical properties of cross-linked and acetylated wheat starches. *Starch/Stärke* 57: 413–420.

Wrigley, C. W. 2016. An overview of the family of cereal grains prominent in world agriculture. In: H. Corke, J. Faubion, K. Seetharaman and C. Wrigley (Eds.) *Encyclopedia of food grains: The world of food grains*, Vol. 1 Oxford, UK: Elsevier, pp. 73–85.

Yang, X., Wan, Z., Perry, L., Lu, H., Wang, Q., Zhao, C., Li, J., Xie, F., Yu, J., Cui, T., and Wang, T. 2012. Early millet use in northern China. *Proceedings of the National Academy of Sciences* 109(10): 3726–3730.

# 2 Shelf Life Enhancement of Pearl Millet Flour

*Shamandeep Kaur, Maninder Kaur, Kawaljit Singh Sandhu and Sneh Punia*

## CONTENTS

## 2.1 INTRODUCTION

Pearl millet (*Pennisetum glaucum*) is a versatile cereal crop, and has a variety of applications in the food and nutrition sector. Pearl millet can grow under challenging environmental conditions and at low cost, where other cereal crops fail to produce an economic yield. Pearl millet flour is a rich source of energy, fat, dietary fiber, mineral composition, high quality protein, and phenolic compounds (Obilana and Manyasa, 2002; Salar and Purewal, 2017). So it is a conventional source of energy and nutrition for the poorest people in the areas where it is grown. In addition to its nutritional benefits, pearl millet has many health benefits, such as possibly helping to prevent certain cancers, lowering blood pressure and cholesterol, delaying gastric emptying, lowering the risk of cardiac diseases, celiac diseases, constipation, diabetes and ulcers due to its high fiber content and mineral components (Nambiar et al., 2011). Despite having a good nutritional profile and health benefits, the full potential of pearl millet flour is limited due to its restricted shelf life and poor keeping qualities. Owing to its high fat content and lipase action, pearl millet flour becomes rancid and produces an off flavor after a certain time in storage (Yadav, Kaur et al., 2012).

In recent times, pearl millet flour was produced daily by housewives. But now, with industrialization and the modern advancements of technology, large amounts of pearl millet are ground by industrial mills, which necessitates storage (Kapoor and Kapoor, 1990). The development of rancidity and bitterness in pearl millet flour has been a serious problem in its commercialization for various food products (Kaced and Hoseney, 1984). Urbanization has created a demand for pearl millet flour with a longer shelf life. To try to prevent rancidity and enhance the shelf life of pearl millet flour, pearl millet grain was subjected to various processing methods (Nantanga et al., 2008).

For different purposes, shelf stability of flours is achieved by deactivating the enzymes responsible for initiating the deteriorative reactions. Various processes have been developed to lessen the development of rancidity in the stored flour. These include use of different containers for storage (Dahiya and Kapoor, 1983), defatting (Kapoor and Kapoor, 1990), acid soaking (Chavan and Kachare, 1994), steaming after and before pearling (Yadav, Kaur et al., 2012) and microwave treatment (Yadav, Anand et al., 2012). To prepare various value-added products from pearl millet flour, there is a need to increase the shelf life of the flour by using minimum processing conditions.

## 2.2   FLOUR RANCIDITY

The pearl millet grain is small, with a larger germ part, so it has a higher content of triglycerides. When the grain is milled into flour in the presence of moisture and oxygen, the lipase activity of the flour is increased, which has a deleterious effect on the quality of the flour. The lipase enzyme, which is concentrated in the pericarp, aleurone layer, and germ, accounts for the triglyceride hydrolysis in pearl millet grain, resulting in an off odor and taste in the flour and its products (Galliard, 1999).

Rancidity is the development of unpleasant odors and flavors in foods resulting from deterioration in the fat and oil portion of the food. It is a natural process of decomposition of fats and oils, either by hydrolysis or oxidation, or both. Lipids are prone to hydrolytic and oxidative degradation, which begins when lipases start to deteriorate acyl lipids as soon as the grain structure breaks, which results in the formation of off flavors and bitter taste (Doehlert et al., 2010; Lehtinen and Kaukovirta-Norja, 2011).

Pearl millet has a high lipid content (7–7.9%), which contains a high amount of triglycerides that are rich in unsaturated fatty acids. The pearl

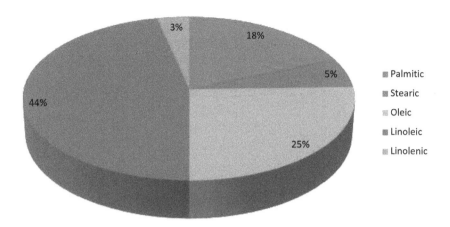

**FIGURE 2.1** Typical fatty acid composition of pearl millet triglycerides (Nantanga, 2006; Rooney, 1978)

millet fat contains 74% of unsaturated fatty acids (oleic, linoleic and lino-lenic) and 26% of saturated fatty acid residues (palmitic and stearic). The fatty acid profile of pearl millet triglycerides is described in Figure 2.1. The development of rancidity in pearl millet flour during storage is, thus, due to various causes such as hydrolysis of lipids, oxidative changes in unsaturated fatty acids, enzymatic changes in C-glycosyllavones, and the presence of phenolics and their enzymatic degradation (Kaced and Hoseney, 1984; Lai and Varriano-Marston, 1980; Reddy et al., 1986) (Figure 2.2).

Oxidative rancidity in foods refers to the perception of objectionable fla-vors and odors caused by oxidation of the unsaturated fatty acid chains of lipids by atmospheric oxygen. Lipid oxidation not only affects the quality of foods by imparting impaired flavors and odors, but also causes loss of essen-tial nutrients such as fatty acids and vitamins, and changes in texture and colour as a consequence of the reaction of lipid oxidation components with other food components (Velasco et al., 2010). Oxidative rancidity of pearl millet oil results in hydroperoxides (chain reaction throughout autoxidation) and, consequently, the production of off flavor generating volatile secondary metabolites (aldehydes, ketones, acids, polymers, etc.).

Hydrolytic rancidity is caused by breaking down of a lipid into its compo-nent fatty acids and glycerol. Enzymatic rancidity or hydrolytic rancidity occurs mainly due to the enzyme lipase. Hydrolytic rancidity is the produc-tion of free fatty acids through the action of lipase and, further, produces the bitter taste and off-odor-causing phenolic aglycones through the action of peroxidase on C-glycosyl flavones (Eskin and Przybylski, 2001). Bitter com-pounds are also formed due to enzymatic browning through the action of polyphenol oxidase (PPO).

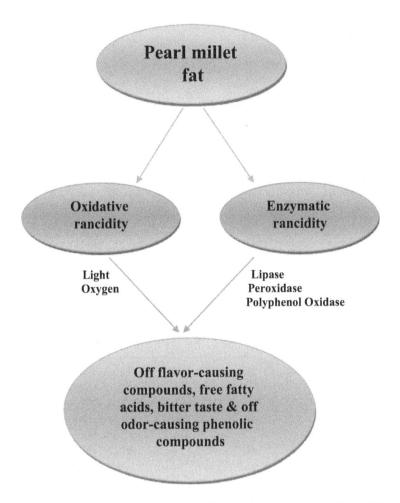

**FIGURE 2.2**   The schematic representation of the mechanism of onset of rancidity (rancidity mechanism, Mazumdar et al., 2016)

## 2.3   VARIOUS METHODS OF STORAGE STABILITY FOR PEARL MILLET FLOUR

Rancidity and bitterness in pearl millet has been a major issue in its development as a commercial product (Kaced and Hoseney, 1984). This has created a demand for pearl millet flours with excellent keeping qualities and a longer shelf life. To prevent rancidity and enhance the storage stability of pearl millet flour, pearl millet grains were subjected to various treatments, such as hydrothermal treatments (steaming, blanching, and boiling), thermal treatments (toasting and microwaving), pearling/debranning, fermentation, use of antioxidants, use of different storage containers, acid soaking, defatting

**TABLE 2.1**

**Treatments used to improve the storage stability of pearl millet flours**

| Name of Technique | Treatments | Conditions (Temperature/Time) | Effect on Properties/Components Present in Flour | References |
|---|---|---|---|---|
| Hydrothermal treatments | **Steaming** Steaming before pearling (SBP, 20 min) Steaming after pearling (SAP, 20 min) | 1.05 kg m$^{-2}$ 0–25 min | • Reduces the lipase activity, free fatty acids and peroxide value of pearl millet flour • Increases the shelf life of flour up to 50 days | Yadav, Kaur et al., 2012 |
| | **Boiling** Boiling in water | 15 min | • Fat acidity value remains stable in boiled sample whereas it increases fivefold in the untreated sample | Chavan and Kachare, 1994; Nantanga et al., 2008 |
| | **Blanching** Blanching in hot water | 98°C 10–30 sec | • Retards the enzymatic activity and development of fat acidity in pearl millet flour | Chavan and Kachare, 1994; Kadlag et al., 1995; |
| Thermal treatments | **Toasting** Oven heating | 100°C 30–120 min | • Decreases the acid value, fat acidity and % FFA 3–4 fold | Arora et al., 2002; Kadlag et al., 1995; Nantanga et al., 2008 |
| | **Microwave treatment** | (900 W, 2,450 MHz for 40–100 sec) | • Decreases lipase activity • Reduces the FFA value and peroxide value • Reduces the level of anti-nutritional compounds | Yadav, Anand et al., 2012 |
| Fermentation | | 36 h | • Reduces the level of fat acidity and phytic acid content | Tiwari et al., 2014 |

(Continued)

**TABLE 2.1 (Cont.)**

| Name of Technique | Treatments | Conditions (Temperature/Time) | Effect on Properties/ Components Present in Flour | References |
|---|---|---|---|---|
| Addition of antioxidants | BHA (0.02%), BHT (0.02%) and ascorbic acid (0.5%) | | • BHA retards hydrolytic and oxidative decomposition of fats<br>• Controls phytic acid and peroxide values<br>• BHT and ascorbic acid also reduce the level of fat acidity and peroxide value | Kapoor and Kapoor, 1990 |
| Defatting | Keeping flour in n-hexane with occasional shaking | 12 h<br>Hexane:Flour 10:1 | • Reduces the fat acidity and peroxide values | Kapoor and Kapoor, 1990 |
| Pearling/Debranning | Pearling by using rice polisher to remove the outer pericarp layer | 5–30 min | • Decreases the level of phytic acid, at acidity and FFA value of pearl millet flour | Tiwari et al., 2014 |
| Acid soaking | Hydrochloric acid solution | 0.05 M | • Inhibits the hydrolysis of triglycerides | Chavan and Kachare, 1994; Rai et al., 2008 |
| Using different packaging materials | Polythene bags, plastic box and cotton bags | | • Lipid deterioration was maximum in cotton bags while lower in polythene bags and plastic box | Kaced and Hoseney, 1984; Kadlag et al., 1995 |

process of flours, etc. Different techniques used to increase storage stability of flours are presented in Table 2.1.

### 2.3.1 HYDROTHERMAL TREATMENTS

Hydrothermal treatments are processing methods used to improve the processing quality and enhance the storage stability of flours. It is considered to prevent the deterioration of triglycerides and deactivate the lipase enzyme in pearl millet flour. Hydrothermal methods are used to process the flour using high temperatures in the presence of water. Hydrothermal treatments had higher lipase inhibition capacity, as compared to thermal treatments, because wet heat has a higher specific enthalpy than dry heat (Nantanga et al., 2008). Three processing methods of hydrothermal treatments have been given to pearl millet grains to increase the quality of flour and storage stability by reducing the rancidity. A flow chart describing processes of hydrothermal treatments is presented in Figure 2.3.

#### 2.3.1.1 Steaming

Steaming is an effective method to extend the shelf life of pearl millet flour. The steaming of grains has been done in a laboratory autoclave for 15–25 min at 1.05 kg cm$^{-2}$ followed by drying (50 ± 5°C in a mechanical tray oven. The impact of steaming on the whole and pearled pearl millet grains confirmed that flour from steaming before pearling (SBP, 20 min) and flour from steaming after pearling (SAP, 15 min) has no lipase action and is observed to have suitable physical, functional, and pasting properties, and an increased storage life of up to 50 days under ambient (15–35°C) temperatures as compared to untreated flour, which has 10 days' storage stability. Steaming lowers the lipase activity and peroxide value of pearl millet flours. The lipase activity of pearl millet flour reduced considerably with an increase in the duration of steaming. It was also observed that the free fatty acid value increased rapidly in untreated samples as compared to treated samples. Increase in FFA values was reported through development of rancidity that was attributed to auto-oxidation deterioration. Steaming of grains also reduces the level of total phenols and tannin content of pearl millet flour, which are directly responsible for bitterness in flour, and also had an effect on color (Yadav, Kaur et al., 2012).

#### 2.3.1.2 Boiling

Boiling is one of the hydrothermal treatments used to increase the storage stability and keeping qualities of pearl millet flours. Boiling water treatment of grain is simple, economical, and adaptable for both domestic and large-scale processing. Pearl millet grains are boiled in a pan of boiling water for 15 min. Thereafter, grain is spread on trays and placed in an oven (at 40°C) to dry for 24 h. The fat acidity of flour from the untreated grain increased from 0.11 to 3.73g KOH/kg during three months of storage, whereas no major increase was observed for wet thermal treated samples. This indicates that no fat acidity develops in boiled pearl millet grain flours (Nantanga et al., 2008). Fat acidity

**Hydrothermal Treatments**

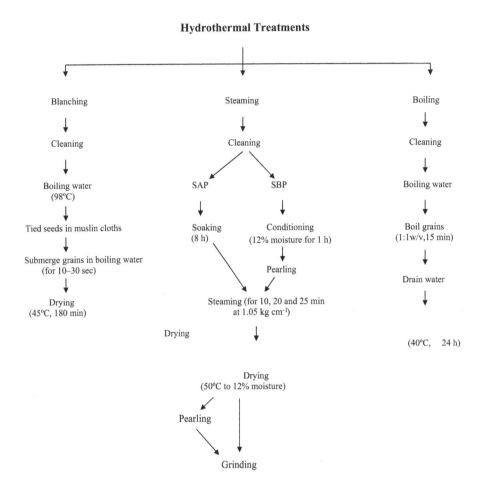

**FIGURE 2.3** Flow chart showing different hydrothermal treatments used to increase shelf life of flours

increased over fivefold in untreated and hot air treated samples, whereas it remained almost unchanged in the flour of the boiled grain during storage in normal conditions for a month (Chavan and Kachare, 1994).

### 2.3.1.3 Blanching

Blanching is an individual heat treatment designed to deactivate enzymes. It is also a common operation prior to freezing, canning, or drying, in which the food samples are heated for the purpose of deactivating enzymes, modifying texture, preserving color, flavor, and nutritional value, and removing trapped air (Fellows, 2000). The process involves the seeds being tied in muslin cloth and dipped in boiling water for 10 or 20 s, drained, then dried at 40°C in a hot air oven. Blanching is one of the impressive techniques for enhancing the shelf life of pearl millet flour. It is used to slow down the rate of enzymatic reactions without having any significant effect on nutritional composition. Short time blanching is effective in reducing the incidence of deterioration reactions during storage. Blanching of seeds at 98°C for 10–30 s before processing has been found to successfully inhibit the lipase activity, free fatty acid value, and the development of fat acidity in meal. It has been proved to be effective in the retardation of enzymatic activity and, thus, improves the shelf life of pearl millet flour by about a month at ambient temperatures without altering the dietary profile much (Chavan and Kachare, 1994; Kadlag et al., 1995). It is found that unpleasant changes in flour can effectively be lessened by a simple hot water blanching treatment of the grains. Blanching of grains for 10–20 s has been as effective as dry heating of grains for 120 min. The value of fat acidity and percentage of FFA of flour obtained from blanched grains showed a three- to fourfold decrease when compared with the flour from untreated grains (Kadlag et al., 1995). The fat acidity of pearl millet flours increased over six times in a fresh sample, but it remained nearly unchanged in flour obtained from blanched grains (Chavan and Kachare, 1994).

## 2.3.2 Thermal Treatment/Dry Heat Treatments

Lipase activity is the leading cause of deterioration of pearl millet flour, so its inactivation before milling improves the quality of product. The application of dry heat to pearl millet meal effectively retards lipase activity and minimizes lipid breakdown during storage (Rai et al., 2008). Thermal processing is the most used method in food preservation to destroy microorganisms, thus extending storage stability (Randhir et al., 2008). Microwave treatment and the hot air oven method are dry heat treatments used to enhance the shelf life and keeping qualities of pearl millet grains. A flow chart describes the process of thermal treatments (Figure 2.4).

### 2.3.2.1 Microwave Treatment

The microwave technique is a method that is significantly attractive when carried out generally in food production (Venkatesh and Raghavan, 2004). Microwave energy is a non-ionising form of power interacts with polar molecules and charged particles of the penetrated medium to cause heat. Microwave heating has the advantage of saving time and energy, enhancing both nutritional quality and the acceptability of some foods by consumers (Nijhuis et al., 1998; Sumnu, 2001).

Microwave treatment can efficiently decelerate the lipase activity, control the FFA value and peroxide value of pearl millet flours, and enhance the shelf life of flour in ideal situations. Microwave heating of pearl millet grain changes its FFA value from 20.80 to 22.25% as compared to untreated flour, which has an FFA value from 20.11 to 32.43%. Microwave heating of grains lowers the lipase activity significantly, with the growth in moisture content from 12–18%, and makes its flour palatable for up to 30 days of storage at normal temperatures (15–30°C), whereas the unheated flour has a storage life of only ten days.

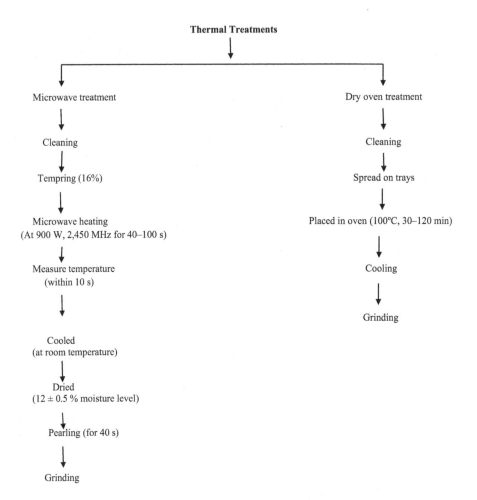

**FIGURE 2.4** Flow chart showing different dry heat treatments used to increase the shelf life of flours

The peroxide values of microwaved grains in flour show no significant changes throughout storage (Yadav, Anand et al., 2012). Similar effects have been recorded on polyphenol oxidase (PPO) enzyme activity in wheat grains treated with microwaves for various time periods at different degrees of grain moisture (Yadav et al., 2008). Microwave heating is faster and provides selective heating ability in contrast to other conventional heating techniques; consequently, the microwave method may be used for lipase deactivation in pearl millet flour (Yadav, Anand et al., 2012).

### 2.3.2.2 Hot Air Oven Method/Toasting

Hot air oven treatment to pearl millet grains at $100\pm2^\circ$C for 30–120 min has been effective for maximum retardation of the lipolytic degradation of lipids and partial deactivation of lipase responsible for the hydrolysis of tri-glycerides during storage (Kadlag et al., 1995; Kapoor and Kapoor, 1990; Rai et al., 2008). The rate of increase in fat acidity, acid value, and per-centage of FFA in heated grain flours was three- to fourfold lower than the untreated grain flours. Dry heating of grains for 120 min was found to be most effective for maximum retardation of the lipolytic decomposition of the lipids during storage (Kadlag et al., 1995). The fat acidity of flour from the untreated grain increased from 0.11 to 3.73g KOH/kg during three months of storage, whereas it increased from 0.01 to 0.69g/kg in a thermal treated flour sample (Nantanga et al., 2008). Heating of grains in a dry oven at 100°C for 2 h reduces the extent of hydrolytic rancidity of the resulting flour during one month's storage in ambient conditions. Fat acid-ity and FFA values increase during storage, though heat treated grain flour increased by less than twofold while the content in untreated grain flour increased at least fourfold. Reduction of free fatty acid content after dry heating might result from the deactivation of lipase activity at high temper-atures. Chapattis made with flour from heat-treated grains remained palat-able during one month's storage without any adverse effect, whereas those made with the untreated grain flour were not acceptable after ten days (Arora et al., 2002). Thermal treatment of grains also showed a substantial reduction in anti-nutritional compounds such as polyphenols (from 3 to 2.27 g/100 g) and tannins (from 1.52 to 1.30 g/100 g) (Nithya et al., 2007).

### 2.3.3  FERMENTATION

Fermentation is a basic processing method for food preservation, flavor devel-opment, and for enhancement of the nutritional quality of food products (Saleh et al., 2013). The fat acidity in fermented grain flour is stable up to the fourth day of storage, whereas it increases on the first day of storage in the untreated flour. This indicates hydrolytic decomposition as well as oxida-tive degradation (Kaced and Hoseney, 1984; Tiwari et al., 2014). Fermenta-tion of grains for 36 h also reduces the phytic acid level from 728 mg/100 g to 398 mg/100 g. The phytic acid content decreases with an increase in fer-mentation time (Tiwari et al., 2014). Fermentation also improves the *in-vitro* protein digestibility and, hence, the protein quality and availability of pearl millet cultivars (Ali et al., 2003). Fermented cereals have better nutritional quality due to the increased level and bioavailability of nutrients (Gazzaz et al., 1989). A flow chart describes the process of the fermentation technique (Figure 2.5).

### 2.3.4  ADDITION OF ANTIOXIDANTS

Antioxidants are chemical compounds that are capable of slowing or prevent-ing the oxidation of other molecules by retarding the autoxidation of trigly-crides and neutralizing the free radicals (Esan et al., 2018). In food systems, antioxidants are used to prevent lipid oxidation, which results in spoilage,

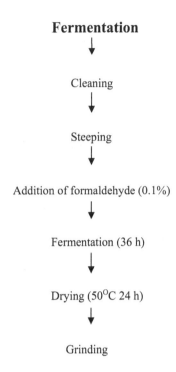

**Fermentation**

Cleaning

Steeping

Addition of formaldehyde (0.1%)

Fermentation (36 h)

Drying (50°C 24 h)

Grinding

**FIGURE 2.5**   Flow chart showing fermentation process used to increase shelf life of flours

quality deterioration, and production of toxic compounds. Antioxidants render their protective effect against oxidative stress via different modes of action, and some may act through a combination of several mechanisms (Shahidi and Zhong, 2010).

Three antioxidants, butylated hydroxyanisole (BHA-0.02%), butylated hydroxytoluene (BHT-0.02%), and ascorbic acid (0.5%), when added to the pearl millet flour, enhance its keeping qualities. The required amount of the chosen antioxidant is first mixed into a small quantity of flour and then this mixture is added to the bulk of the flour, and then passed through a sieve several times to ensure uniform distribution. BHA treatment is able to retard both hydrolytic and oxidative decomposition. The development of fat acidity and peroxide values in an antioxidant treated flour sample was significantly less than in untreated flour samples and the best results were observed in BHA-treated samples (Kapoor and Kapoor, 1990). The lower fat acidity value in BHA treated flour, compared with that of untreated flour, may be due to its antioxidant action. BHT and ascorbic acid also reduce the level of fat acidity and peroxide value as compared to untreated flours.

### 2.3.5 DEFATTING

Defatting is an acceptable technique for enhancing the shelf life of pearl millet flour. This method reduces the fat acidity and peroxide values during storage of pearl millet flour for a month. The peroxide values and fat acidity of defatted flour is about 18 mg/kg and 25 mg KOH/100 g and it was constant during storage for up to 30 days (Kapoor and Kapoor, 1990). A flow chart describing the process of the defatting method is presented in Figure 2.6.

### 2.3.6 PEARLING/DEBRANNING

Pearling, or debranning, is the technique of controlled removal of the outer layers of cereal kernels by abrasion and friction to a desired level prior to milling, simplifying the milling process since less bran remains in the kernel to be removed during milling. The action of removing the outer layers of the grain through debranning is called pearling (Mousia et al., 2004). Pearling is one of the best pre-treatments to enhance the storage stability of pearl millet flour. In a study, pearling of seeds was done for varying duration (5–30 min),

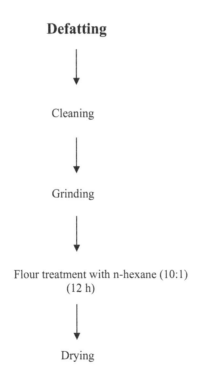

**Defatting**

Cleaning

Grinding

Flour treatment with n-hexane (10:1)
(12 h)

Drying

**FIGURE 2.6** Flow chart showing defatting treatment used to increase shelf life of flours

using a laboratory rice polisher consisting of an abrasive emery roller to remove the outer pericarp layer of seeds. Pearling for 5–30 min was shown to decrease the phytic acid content, fat acidity, and FFA of pearl millet grains, resulting in a maximum 29.5% reduction of phytic acid in pearl millet flour (Tiwari et al., 2014). This is due to the removal of the outer layer of grains, where phytic acid is thought to be abundant. Thus, fat acidity and the FFA of grains is decreased by using pearling of the grain.

### 2.3.7   ACID SOAKING

The acid soaking of pearl millet grains is an effective means to lower pH and lipase enzyme activity. The pearl millet grains are soaked in a 0.05 M hydrochloric acid solution for 12 h at ambient temperature. Then the soaked grains are washed with tap water and dried to 10% moisture content. The fat acidity of acid-soaked grain flour increased 1.5-fold after 30 days' storage, whereas fat acidity of untreated grain flour increased sixfold. This indicated that the acid soaking treatment of grains inhibits the hydrolysis of triglycerides (Chavan and Kachare, 1994). Soaking pearl millet grains in 0.2 N HCL overnight reduced levels of polyphnols and phytic acid by 76% and 82%, respectively. Fat acidity, free fatty acids, and lipase activity increased fourfold in untreated grain flour, while in acid-treated flour it remained constant for up to 28 days of storage (Rai et al., 2008). A flow chart describing the process of the acid-soaking technique is shown in Figure 2.7.

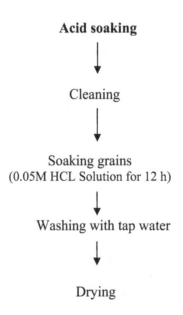

**Acid soaking**

Cleaning

Soaking grains
(0.05M HCL Solution for 12 h)

Washing with tap water

Drying

**FIGURE 2.7**   Flow chart showing acid-soaking treatments used to increase the shelf life of flours

## 2.3.8 Use of Different Packaging Materials

Packaging material has been significant because it prevents the food product from gaining moisture, protects the product against oxygen, and improves the shelf life of food by providing greater safety from all the hazards that might be encountered during storage (Peter and Axtell, 1993). The different storage containers have a significant effect on the lipolytic and oxidative decomposition of fats. Lipid deterioration was highest when the product was stored in cotton bags, while lower levels of change were found when it was stored in plastic boxes or polythene bags under the same conditions (Kadlag et al., 1995). Thus, polythene bags could be preferred for storing pearl millet flour because fat acidity increased more rapidly in cotton bags as compared to polythene bags. The fat acidity for flours stored in cotton and polythene bags increased from 0.4 to 2.2 and 1.5 g KOH kg$^{-1}$, respectively. This effect showed that the cotton bags allowed moisture entry, which hastens the lipase activity and increases fat acidity (Kaced and Hoseney, 1984). Polythene bags can effectively retard moisture entry into the product and increase its shelf life by controlling the hydrolytic rancidity.

## 2.4 CONCLUSION

Pearl millet is gaining importance as a climate-resilient and health-promoting nutritious crop. It is a rich source of energy and nutrition for poor people. Pearl millet flour had a severe problem during storage as it was observed to produce off-flavor and a bitter taste. It becomes rancid and deteriorates after a time during storage due to its high fat content and fatty acid profile. Toward enhancing the storage stability and utilization of pearl millet, various treatments have been shown to be efficient. Hydrothermal and thermal treatments can be used to inhibit hydrolytic rancidity and enhance the shelf life of flours up to 50 days. Pre-milling treatments (pearling and fermentation) reduce the phytic acid content in pearl millet flour. BHA treatment has been able to retard both hydrolytic and oxidative rancidity, which lowers the value of fat acidity and peroxide. The defatting process enhances the storage life of flours up to 30 days without deterioration in keeping qualities. The type of storage containers used for flours has a significant effect on shelf life and keeping qualities.

## REFERENCES

Ali, M. A. M., Tinay, A. H. E. and Abdalla, A. H. 2003. Effect of fermentation on the in vitro protein digestibility of pearl millet. *Food Chemistry* 80: 51–54.

Arora, P., Sehgal, S. and Kawatra, A. 2002. The role of dry heat treatment in improving the shelf life of pearl millet flour. *Nutrition and Health* 16: 331–336.

Chavan, J. K. and Kachare, D. P. 1994. Effect of seed treatment on lipolytic deterioration of pearl-millet flour during storage. *Journal of Food Science and Technology-Mysore* 31: 80–81.

Dahiya, S. and Kapoor, A. C. 1983. Effect of storage conditions on the protein quality of pearl millet flour. *Nutrition Reports International (USA)* 28: 1351–1359.

Doehlert, D. C., Angelikousis, S. and Vick, B. 2010. Accumulation of oxygenated fatty acids in oat lipids during storage. *Cereal Chemistry* 87: 532–537.

Esan, Y. O., Sade, O. O., Victor, E. N. and Oluranti, O. O. 2018. Functional and antioxidant properties of raw and popped Amaranth (Amaranthus cruentus) seeds flour. *Annals Food Science and Technology* 19: 399–408.

Eskin, N. A. M. and Przybylski, R. 2001. Antioxidants and shelf life of foods, in *Food shelf life stability: chemical, biochemical and microbiological changes*, eds. N. A. M. Eskin and D. S. Robinson, 175–209. CRC Press, Boca Raton, Florida.

Fellows, P. J. 2000. *Dehydration-effects on food, in food processing technology* (2nd ed.), 312–340. Woodhead Publishing, Cambridge.

Galliard, T. 1999. Rancidity in cereal products, in *Rancidity in foods*, eds. J. C. Allen and R. J. Hamilton, 140–156. Aspen Publishers, Gaithhersburg.

Gazzaz, S. S., Rasco, B. A., Dong, F. M. and Borhan, M. 1989. Effects of processing on the thiamin, riboflavin, and vitamin b-12 content of fermented whole grain cereal products. *Journal of Food Processing and Preservation* 13: 321–334.

Kaced, I. and Hoseney, R. C. 1984. Factors affecting rancidity in ground pearl millet (Pennisetum americanum L. Leeke) [during storage]. *Cereal Chemistry* 61: 187–192.

Kadlag, R. V., Chavan, J. K. and Kachare, D. P. 1995. Effects of seed treatments and storage on the changes in lipids of pearl millet meal. *Plant Foods for Human Nutrition* 47: 279–285.

Kapoor, R. and Kapoor, A. C. 1990. Effect of different treatments on keeping quality of pearl millet flour. *Food Chemistry* 35: 277–286.

Lai, C. C. and Varriano-Marston, E. 1980. Changes in pearl millet meal during storage. *Cereal Chemistry* 57: 275–277.

Lehtinen, P. and Kaukovirta-Norja, A. 2011. Oats: chemistry and technology, in *Oat lipids, enzymes, and quality*, eds. F. H. Webster and P. J. Wood, 143–156. American Association of Cereal Chemists, Inc (AACC), St Paul, MN.

Mazumdar, S. D., Gupta, S. K., Banerjee, R., Gite, S., Durgalla, P. and Bagade, P. 2016. Determination of variability in rancidity profile of select commercial pearl millet varieties/hybrids. Poster presented in CGIAR Research Program on Dryland Cereals Review Meeting held at Hyderabad, India, 5–6 October 2016. International Crops Research Institute for the Semi-Arid Tropics, Patancheru, Telengana, India.

Mousia, Z., Edherly, S., Pandiella, S. S. and Webb, C. 2004. Effect of wheat pearling on flour quality. *Food Research International* 37: 449–459.

Nambiar, V. S., Dhaduk, J. J., Sareen, N., Shahu, T. and Desai, R. 2011. Potential functional implications of pearl millet (Pennisetum glaucum) in health and disease. *Journal of Applied Pharmaceutical Science* 10: 62.

Nantanga, K. K. M. 2006. *Lipid stabilisation and partial pre-cooking of pearl millet by thermal treatments*. PhD diss., University of Pretoria.

Nantanga, K. K. M., Seetharaman, K., de Kock, H. L. and Taylor, J. R. N. 2008. Thermal treatments to partially pre-cook and improve the shelf-life of whole pearl millet flour. *Journal of the Science of Food and Agriculture* 88: 1892–1899.

Nijhuis, H. H., Torringa, H. M., Muresan, S., Yuksel, D., Leguijt, C. and Kloek, W. 1998. Approaches to improving the quality of dried fruit and vegetables. *Trends in Food Science & Technology* 9: 13–20.

Nithya, K. S., Ramachandramurty, B. and Krishnamoorthy, V. V. 2007. Effect of processing methods on nutritional and anti-nutritional qualities of hybrid (COHCU-8) and traditional (CO7) pearl millet varieties of India. *Journal of Biological Sciences* 7: 643–647.

Obilana, A. B. and Manyasa, E. 2002. Millets, in *Pseudocereals and less common cereals*, eds. S. B. Peter and J. R. N. Taylor, 177–217. Springer, Berlin and Heidelberg.

Peter, F. and Axtell, B. 1993. *Appropriate food packaging*. Publ. Transfer of Technology for Development-Amsterdam, International Labour Office, Geneva.

Rai, K. N., Gowda, C. L. L., Reddy, B. V. S. and Sehgal, S. 2008. Adaptation and potential uses of sorghum and pearl millet in alternative and health foods. *Comprehensive Reviews in Food Science and Food Safety* 7: 320–396.

Randhir, R., Kwon, Y. I. and Shetty, K. 2008. Effect of thermal processing on phenolics, antioxidant activity and health-relevant functionality of select grain sprouts and seedlings. *Innovative Food Science & Emerging Technologies* 9: 355–364.

Reddy, V. P., Faubion, J. M. and Hoseney, R. C. 1986. Odor generation in ground, stored pearl millet. *Cereal Chemistry* 63: 403–406.

Rooney, L. W. 1978. Sorghum and pearl millet lipids. *Cereal Chemistry* 55: 584–590.

Salar, R. K. and Purewal, S. S. 2017. Phenolic content, antioxidant potential and DNA damage protection of pearl millet (Pennisetum glaucum) cultivars of North Indian region. *Journal of Food Measurement and Characterization* 11: 126–133.

Saleh, A. S. M., Zhang, Q., Chen, J. and Shen, Q. 2013. Millet grains: nutritional quality, processing, and potential health benefits. *Comprehensive Reviews in Food Science and Food Safety* 12: 281–295.

Shahidi, F. and Zhong, Y. 2010. Lipid oxidation and improving the oxidative stability. *Chemical Society Reviews* 39: 4067–4079.

Sumnu, G. 2001. A review on microwave baking of foods. *International Journal of Food Science & Technology* 36: 117–127.

Tiwari, A., Jha, S. K., Pal, R. K., Sethi, S. and Krishan, L. 2014. Effect of pre-milling treatments on storage stability of pearl millet flour. *Journal of Food Processing and Preservation* 38: 1215–1223.

Velasco, J., Dobarganes, C. and Marquez-Ruiz, G. 2010. Oxidative rancidity in foods and food quality, in *Chemical deterioration and physical instability of food and beverages*, eds. L. H. Skibsted, J. Risbo and M. L. Andere, 3–32. Woodhead Publishing Limited, New York.

Venkatesh, M. S. and Raghavan, G. S. V. 2004. An overview of microwave processing and dielectric properties of agri-food materials. *Biosystems Engineering* 88: 1–18.

Yadav, D. N., Anand, T., Kaur, J. and Singh, A. K. 2012. Improved storage stability of pearl millet flour through microwave treatment. *Agricultural Research* 1: 399–404.

Yadav, D. N., Kaur, J., Anand, T. and Singh, A. K. 2012. Storage stability and pasting properties of hydrothermally treated pearl millet flour. *International Journal of Food Science & Technology* 47: 2532–2537.

Yadav, D. N., Patki, P. E., Sharma, G. K. and Bawa, A. S. 2008. Effect of microwave heating of wheat grains on the browning of dough and quality of chapattis. *International Journal of Food Science & Technology* 43: 1217–1225.

# 3 Phytochemicals and Antioxidant Properties in Pearl Millet
## A Cereal Grain with Potential Applications

*Kawaljit Singh Sandhu, Pinderpal Kaur, Anil Kumar Siroha and Sukhvinder Singh Purewal*

**CONTENTS**

## 3.1  INTRODUCTION

Detection of phytochemicals with antioxidant properties in natural resources is one of the current focuses of a number of scientists and researchers. Cereal grains are well known for their specific nutrients and bioactive constituents. In addition to the presence of minerals and bioactive compounds, extracts of cereal grains are reported to offer good DNA protection (Kaur, Purewal, Sandhu and Kaur, 2019; Purewal et al., 2019; Singh, Kaur et al., 2019). Bioactive compounds as well as phytochemicals are naturally occurring specific compounds which are formed within the plant system in response to stress conditions (Gan et al., 2019). They are formed in natural resources as secondary metabolites. Because of the presence of certain specific compounds within cereal grain, products based on their flour are recommended for inclusion as an important part of a breakfast menu (Isaksson et al., 2012; Lattimore et al., 2010; Lin et al., 2017). Although each cereal grain has its own importance, millet grains, with their drought resistant nature, nutritional profile and certain specific health benefits, attract the interest of scientists throughout the world (Kaur, Purewal, Sandhu, Kaur and Salar, 2019; Salar and Purewal, 2016; Salar et al., 2017a; Saleh et al., 2013). The amount and presence of specific phytochemicals in cereal grains may vary from cultivar to cultivar (Salar and Purewal, 2017). Phytochemicals with antioxidants potential are reported to possess health benefiting properties. Diseases and abnormality in functioning of specific body parts is one of the leading causes of death worldwide. Diseased conditions may be due to genetic defects, consumption of unhealthy food, excessive use of junk foods, a person's age, life style, and eating habits, and environmental factors. Cases of persons suffering from diabetes are increasing day by day in both developing and developed countries. Scientists throughout the world are working on different aspects of natural resources to combat various diseases. Extracts from natural resources are known to combat oxidative stress, with many more health benefiting properties. Nani et al. (2011) studied the effect of pearl millet consumption on rats suffering from diabetes. They observed that meal based on pearl millet has an important role in curing hyperglycemic conditions in diabetic patients. The presence of specific bioactive compounds in pearl millet is associated with the reduction of HT29 tumor cells (Chandrasekara and Shahidi, 2011).

Extraction of phytochemicals from natural resources is done by converting them into a suitable liquid medium, and is condition dependent, which is the main reason for their varying amounts. The selection of a suitable extraction phase and extraction conditions are recommended for optimal extraction of phytochemicals from natural resources (Cheok et al., 2012; Liyana-Pathirana and Shahidi, 2005). Extraction conditions are one of the critical factors that need to be studied in detail while working on phytochemicals of natural origin (Casagrande et al., 2018; Irakli et al., 2018; Oussaid et al., 2017; Salar et al., 2016; Xiong et al., 2019).

An increasing demand for food with health benefiting properties means that the researchers need to pick the natural resources with functional properties to work on. To meet the increasing hunger rate in developing nations, where the increase in population is very high requires the discovery of natural resources that could act as a source of food for both humans and animals. Millet crops are important as they can tolerate dry growing conditions. India is a leading nation in terms of millet production, especially that of pearl millet (Kaur, Purewal, Sandhu, Kaur and Salar, 2019). Products based on pearl millet flour have lower gluten content, and, therefore, could be recommended to persons with a gluten allergy. Various products of pearl millet are now available in the market and they are well known for their taste as well as their nutritional benefits (Adebiyi et al., 2017; Bora et al., 2019; Kumar et al., 2018, 2019; Wang et al., 2019).

Like other natural resources, pearl millet also contains phytochemicals with antioxidant properties; however, they are present in a complex form and need to be converted to a simpler form in order to maximize their benefits (Salar and Purewal, 2016). Solid state fermentation is an important biotechnological technique adopted by food scientists for the enhancement and improvement of specific phytochemicals within the natural resources (Bei et al., 2018; Salar et al., 2012; Shin et al., 2019). Various biochemical changes occur in the natural resource while being processed under different sets of conditions. This chapter focuses on phytochemicals and antioxidant properties in pearl millet.

## 3.2 AGRICULTURAL AND NUTRITIONAL IMPORTANCE OF PEARL MILLET

Pearl millet is an important cereal grain from an agricultural point of view as it requires only a limited amount of water for its growth and has a short growing period. Even under quite harsh environmental conditions, a pearl millet crop offers reliable harvesting (Salar and Purewal, 2017; Siroha et al., 2016). Grains of the pearl millet crop are highly nutritious and easily available in the market at very low prices. Pearl millet is a Kharif crop with small-sized grains and many health benefits. The amount of fertilizers required to grow a pearl millet crop is less than that needed for other cereal grains. The adaptive nature of a pearl millet crop provides gives it special status when compared with other cereal grains. Being a C4 plant, pearl millet is photo-synthetically highly active and, thus, capable of producing a high yield of dry matter (Yadav and Rai, 2013). A linkage map of pearl millet with molecular markers was prepared with restriction fragment-length polymorphism (RFLP) (Liu et al., 1994). The preparation of such genetic maps helps researchers to study the various genetic characteristics present in pearl millet (Busso et al., 2000; Gale et al., 2005).

Pearl millet grains are rich in macronutrients as well as micronutrients, which are required for a healthy lifestyle (Figure 3.1). Pearl millet flour is being used in the preparation of various traditional dishes and bakery products. Depending on the specific type of millet, the amount of minerals and other constituents may vary. Scientific reports indicate the presence of nutritionally important constituents in millets such as iron, with a range of 0.02–0.01 g kg$^{-1}$; calcium (0.1–3 g kg$^{-1}$); phosphorus (2–3.4 g kg$^{-1}$); zinc (0.3–0.6 g kg$^{-1}$); and the vitamins niacin (0.9–11.1 mg kg$^{-1}$), riboflavin (2.8–16 mg kg$^{-1}$) and thiamine (1.5–6 mg kg$^{-1}$). The energy values of pearl millet based functional food products are 1500–1700 KJ 100 g$^{-1}$. Like other important cereal grains, pearl millet also contains a varying amount of starch as a major carbohydrate presence in grains. The amount of starch present in pearl millet grains varies from 60–72 g 100 g$^{-1}$. The difference in the amount of starch may be attributable to the smaller size of pearl millet grains as compared to other cereals (Kaur, Purewal, Sandhu, Kaur and Salar, 2019; Sandhu and Siroha, 2017). The amylose content of pearl millet starch may vary from 15–22%, depending on the cultivar and starch processing methods. Pearl millet grains are a good source of protein (10–15%) as well as fibers (6–10%). Being a source of protein and fibers, pearl millet grain flour could be used in the preparation of various functional food products and weight management therapies (Dias-Martins et al., 2018; Shaikh et al., 2019). The gluten-free nature of pearl millet flour could be a boon for patients suffering from celiac disease. The ash content in pearl millet cultivars may vary from 1–3.5%, depending on their anatomical structure as well as the environment in which they have grown. Various gluten-free products could be prepared from pearl millet flour.

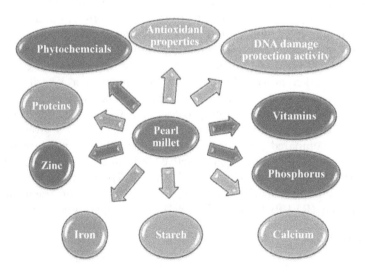

**FIGURE 3.1**   Important nutritional and health benefiting components of pearl millet

The presence of a high fat content (5–7%) in pearl millet results in a shorter shelf life as compared to other cereal grains with a low fat content. A high amount of fat content causes oxidative rancidity.

Pearl millet grains are also a good source of vitamins E, B and A. The mineral profile of pearl millet grains is highly affected during processing methods (milling), as the minerals are present mainly in the germ, pericarp and aleurone layer.

## 3.3 ANTIOXIDANT PROPERTIES

Phytochemicals present in pearl millet are well known for their antioxidant properties. Compounds with antioxidant properties are useful as a defense mechanism against environmental stress, as well as slowing down the effect of free radicals/oxidative stress. Salar and Purewal (2017) studied 12 pearl millet cultivars and they observed activity/inhibition during different assays DPPH (22.71–54.90%); ABTS (27–52.01%) and HFRSA (0.50–6.11%). They observed maximal activity in the pearl millet cultivar PUSA-415. Siroha et al. (2016) reported the presence of metal chelating activity in six different cultivars of pearl millet, with a range of 6.7–20.3%. From their observation, the best pearl millet cultivar in terms of antioxidant properties is GHB-732. Various antioxidative enzymes are also involved during oxidative stress conditions, and these are ascorbate peroxidase, dehydroascorbate reductase, polyphenol oxidase, superoxide dismutase and catalase (Avasthi et al., 2018; Zhang et al., 2013). Specific phytochemicals present in natural resources have the capability to slow down the damaging effect of oxidative stress (Dhull et al., 2016; Salar et al., 2015). Most of the natural resources possess phytochemicals in their matrix in complex form/bound form, which has to be released from bound to free form so as to initiate maximal activity in the digestive system. Various processing methods adopted by researchers are being applied to natural resources to increase the amount of extractable bioactive phytochemicals. However, despite their bound form, phytochemicals are condition dependent, as their liberation from natural resources depends on the extraction phase type, concentration of solvents used, temperature during the extraction process and time duration (Omwamba and Hu 2009; Salar et al., 2016). Each solvent has its specific effect on the mobility of phenolics and other important constituents. The efficacy may vary with the nutritional composition of the particular natural resource. Phytochemical constituents are solvent specific and experimental conditions have to be set in a manner that could directly affect the bioactive profile.

## 3.4 PEARL MILLET PHYTOCHEMICALS AND THEIR HEALTH BENEFITS

Phytochemicals present in whole grains and their fractions are gaining importance because of the presence in them of health benefiting properties.

Extraction and characterization of phytochemicals from cereal grains mainly depends on the extraction parameters. Utilization of products based on cereal grains is helpful in combating health problems. Scientific reports available on phytochemicals suggests their role in maintaining healthy life style. The major bioactive compounds present in pearl millet and milling fractions are as set out in the following sections.

### 3.4.1   p-COUMARIC ACID

Chandrasekara and Shahidi (2011) reported the presence of p-coumaric acid in millet grains. Rao and Muralikrishna (2001) observed that caffeic acid, ferulic acid and p-coumaric acids were the major bound phenolics in finger millet. p-Coumaric acid is an important polyphenol which is present in grains and responsible for the presence of antimicrobial properties in them. Coumaric acid is widely distributed and present in bound forms. Orhto, meta and para are the three different forms of cinnamic acid which are present in natural resources. p-Coumaric acid is the dominant form widely present in natural resources. Normally, it is present in cell walls and its key function is to decrease the low-density level lipoproteins, free radical quenchers and antimicrobial agents (Boz, 2015; Gani et al., 2012; Roy and Prince, 2013; Yoon et al., 2013).

### 3.4.2   GALLIC ACID

Rao and Muralikrishna (2002) reported that specific bioactive compounds, that is, p-coumaric, gallic and ferulic acids, start increasing after malting. Salar and Purewal (2017) reported the effect of solid state fermentation on gallic acid in pearl millet. The amount of gallic acid was increased after a fermentation duration of six days. Gallic acid is an important bioactive constituent found as a colorless powder present either as a free molecule or in a complex form with tannins. The presence of antioxidant potential, together with antiviral and antifungal properties, in gallic acid makes it a useful compound for pharmaceutical preparations. Consumption of natural resources that are a rich source of gallic acid may boost the immune system and the production of pathogen eliminating leukocytes. Gallic acid also protects the brain from neurodegenerative disorders. Godstime et al. (2014) reported the presence of the inhibitory potential of gallic acid against bacterial dihydrofolate reductase and its excitatory potential on topoisomerase IV-mediated DNA cleavage. Gallic acid and its ester derivatives (octyl gallate) are well known for using the hydrophilic catechol part for binding to the polar face of cellular membranes and they enter the lipid bilayer, which further results in alteration to cellular permeability in fungal strains (Kubo et al., 2003). Gandhi et al. (2014) observed that gallic acid is involved in inhibiting diet induced hypertriglyceridemia and hyperglycemia, which results in reduction of the size of adipocytes and protects pancreatic β-cells by inducing peroxisome proliferator-activated receptor-γ (PPAR-γ) expressions.

### 3.4.3 FERULIC ACID

Kern et al. (2003) observed the absorption of ferulic acid in the small intestine and they reported that ferulic acid was the soluble fraction of bioactive compounds present in cereal grains. They reported that a minute quantity of ferulic acid was absorbed in the large intestine. Ferulic acid, being an important antioxidant in natural resources, offers many health benefits to the consumer. It plays an important role in diabetes control, blood pressure management and protection of skin. Although ferulic acid has health benefiting properties in it,, its effect on living organisms may vary depending on host age, the administration route to the host cell and the immune system.

Ferulic acid is formed in the plants through the shikimate pathway, with L-phenylalanine and L-tyrosine as sole entities (Graf, 1992). Ruan et al. (2008) reported that a high level of ferulic acid is responsible for antioxidation. It also plays a role in the pigmentation of yolk, lipid stability and the thickness of a shell (Freitas et al., 2017; Tesaki et al., 1998). Owing to its hydrogen donating nature, ferulic acid becomes an important free radical quenching compound that further helps the body through anti-inflammatory effects (Ou and Kwok, 2004; Rosa et al., 2013).

### 3.4.4 BENZOIC ACID

Benzoic acid is an important carboxylic acid which has many uses in the cosmetic, pharmaceutical and food industries. In food industries, it is being used as a preservative. Benzoic acid has antifungal properties, helps the body to shed rough and dead cells and protects the body against skin irritations. Benzoic acid is well known as a food preservative, food additive and flavoring agent. Owing to their higher solubility, the derivatives of benzoic acid are often used as antimicrobial agents. The action of benzoic acid involves an alteration in cell membranes, which includes the accumulation of toxic compounds, disturbs the ion equity and alters the pH homeostasis (Brul and Coote, 1999; Iammarino et al., 2011; Krebs et al., 1983). Derivatives of benzoic acid are broad spectrum antimicrobial agents (Heydaryinia et al., 2011). Benzoic acid could also be used as a feed additive and consumption of it in feed results in a reduction in diarrhea and increased body weight in piglets (Papatsiros et al., 2011).

### 3.4.5 ASCORBIC ACID

Malleshi and Klopfenstein (1998) reported the presence of ascorbic acid in pearl millet. Vitamin C (Ascorbic acid) is a water soluble vitamin present in natural resources and provides many health benefits as it is involved as a cofactor in various biochemical reactions. Regular consumption of vitamin C-enriched food and food products may help to boost immunity and maintain the cardiovascular system and healthy skin. The health benefits of ascorbic acid is dose dependent, and the recommended dose in scientific reports is

60–75 mg for females and 75–90 mg for males per day. The dose may vary from person to person, depending on their daily routine, exercise and body metabolism. Consumption of vitamin C in doses of more than 2000 mg per day may result in adverse health effects such as headache, diarrhea and stomach cramps, etc.

### 3.4.6   CINNAMIC ACID

Bioactive phytochemicals are secondary metabolites which are widely distributed in natural resources. They are present in fruit, vegetables, cereal grains and other important plant-based foods. The presence of phytochemicals in natural resources affects taste, texture, color and health benefiting properties (Bento-Silva et al., 2019). Phytochemicals play an important role during the growth of plants and in their specific developmental stages, reproduction and disease resistance (Solecka, 1997; Ververidis et al., 2007). Cinnamic acid is one of the important phytochemicals being used by pharmacists to alter permeability, solubility and potency. Cinnamic acid is present in natural resources in two major forms: free and bound. Free forms are present mostly in fruits and vegetables; the bound form is present in cereal grains. The amount of specific phytochemicals may vary with the type of natural resources, its growth stage, growing conditions, storage conditions and processing methods. Scientific reports suggest the medicinal role of cinnamic acid in a series of biological activities and in the prevention of diseases related to metabolic disorders, cardiovascular abnormality and diabetes (Alam et al., 2016; Loader et al., 2017; Meng et al., 2013).

### 3.4.7   SYRINGIC ACID

Syringic acid possesses important medicinal properties, such as anti-inflammation, antimicrobial, antioxidant and anti-diabetic. Production of syringic acid has been reported in natural resources as well as certain fungal strains. Antimicrobial agents of natural origin are gaining more importance as they pose no risk to health (Joshi et al., 2014). Extracts of syringic acid possess broad spectrum antibacterial properties. The extracts are effective against both gram positive and gram negative bacterial strains (Manuja et al., 2013). Consumption of syringic acid-rich natural resources results in decreased endotoxins and lipo-polysaccharide-induced death in rabbits (Liu et al., 2002).

### 3.4.8   VANILLIC ACID

Vanillic acid is an important example of bioactive compounds, as it has its own importance in maintaining biological reactions. Consumption of vanillic-rich natural resources in diets may helpful in preventing the symptoms of cellular damage. Vanillic acid is well known for DNA repair and slowing down the symptoms of Parkinson's disease.

### 3.4.9 SALICYLIC ACID

Salicylic acid is an important bioactive constituent which is actually a beta hydroxy acid present in natural resources. Salicylic acid is being used in cosmetic industries for the preparation of formulations that can slow down the ageing rate, clear skin pores and treatment of oily skin. Formulations prepared from salicylic acid are helpful in getting rid of blackheads and whiteheads on the face. Consumption of salicylic-rich natural food and food products are helpful to avoid acne.

### 3.4.10 CATECHOL

Catechol is an important bioactive phytochemical which is the predominant product of aerobic catabolism of specific phenolic compounds (Nesvera et al., 2015). Catechol is widely present in natural resources and required for the maintenance of various reactions in the body. Catechol possesses antioxidant properties which could be helpful in lowering the activity of free radicals during oxidative stress conditions (Li et al., 2009; Salar and Purewal, 2016).

## 3.5 BIOTECHNOLOGICAL APPROACHES FOR THE IMPROVEMENT OF PEARL MILLET

Scientific staff/researchers are continuously working on the improvement of the nutrition profile and various other characteristics of cereal grains and natural resources. Scientific studies reported the effect of different processing conditions on the nutritional profile of grains (Sharma and Gujral, 2010). Microbial and molecular biotechnology tools are currently being used either to cause changes/modifications at marker level or at the enzymatic level. Fungal strains-based fermentation of natural substrates, especially cereal grains, is currently on trend. Production of enzymes by fungal strains during the fermentation process induces biochemical changes in the substrate that will help the bioactive compounds to move from their bound form to free form (Postemsky et al., 2017; Salar and Purewal, 2016). Cereal grains and other natural resources are a rich source of bioactive constituents which could be used to prevent/delay the action of free radicals within the body (Singh, Khan et al., 2019; Singh, Kaur et al., 2019). However, the bioactive constituents in natural resources are present in a complex and conjugated form that results in slow utilization in the digestive tract. Free forms of these bioactive constituents are easily absorbed by the intestinal system. The capability of microorganisms to induce transformation of specific compounds at the microbial level makes them a perfect choice for physiological as well as anatomical changes in the natural resources (Salar et al., 2017a). Bacterial strains and fungal strains are being used to modulate the nutritional profile. However, depending on the type of substrate and incubation conditions, the choice of starter culture varies accordingly.

### 3.5.1 FERMENTATION TECHNOLOGY

Microorganisms-based fermentation of natural resources is currently on trend for the improvement and modification of their bioactive profile. Fermentation is an important process which needs to be performed by expert microbiologists to ensure the success of process parameters and objectives. The effect of starter culture on steam-sterilized substrates depends on inoculum size, age of the starter culture, incubation conditions, the nutritional profile of the substrate and extraction conditions (Bhanja et al., 2009; Salar et al., 2017b). Natural resources, under certain circumstances, start producing bioactive phytochemicals which are present in complex forms and, for complete absorption, need to be converted to free form (Salar et al., 2012). Fungal, as well as bacterial, strains are being used as starter culture for the fermentation process. As compared to bacterial fermentation, fungal strains-based fermentation of natural resources shows promising results. The ability to grow within an environment with a limited amount of moisture is one of the important features present in fungal strains. Fungal strains that are filamentous in nature are generally used for the fermentation process, as they have the capability to be incorporated in steam-sterilized soft substrates, to produce specific enzymes and bring out phenolics from their bound forms to their free forms. Production of specific enzymes such as α-amylase; xylanase; β-glucosidase and phytase are responsible for the increased antioxidant properties in fermented substrates.

Microorganisms are being employed to create the transformation of bioactive compounds, as they are capable of performing the transformation process in a shorter duration of time and in a cost-effective manner. *Aspergillus awamori*, *Aspergillus sojae* and *Aspergillus oryzae* was used for the purpose of modulating the bioactive profile of pearl millet cultivars HHB-197 and PUSA-415 (Salar and Purewal, 2016; Salar et al., 2016; Salar et al., 2017a). Fungal fermentation of pearl millet results in the enhancement of bioactive compounds by 3–5fold depending on the starter culture and incubation conditions (Figure 3.2). Purewal et al. (2019) carried out the fermentation of pearl millet with *Rhizopus azygosporus* for ten days and they observed the changes in phenolic compounds by 3.31fold until the last day of the fermentation experiment. Microorganisms incorporate their filaments in steam-sterilized substrate to utilize them as a source of food either for their growth or to maintain the physico-chemical reactions in them. For the purpose of microbial transformation at the desired level, it is better to understand the factors that need to be optimized before fermentation. Typically, the fermentation conditions, extraction parameters and enzymes produced by a starter culture during microbial fermentation are responsible for the enhancement of phenolics and their conversion from bound forms to free forms. Microbial consortia, in response to certain specific conditions, itself acts as source of important mycochemicals and, simply, they incorporate their own mycochemicals in fermented substrates (Salar et al., 2017c).

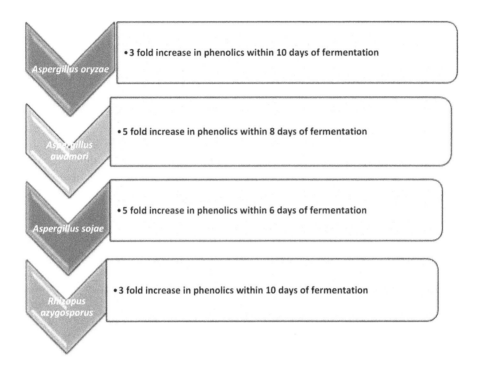

**FIGURE 3.2** Effect of solid state fermentation on pearl millet bioactive profile

## 3.5.2 Disease Resistance

Pearl millet is an important crop that shows a high degree of heterosis for grains as well as stover yields. In the beginning, hybrids were prepared by allowing the growth of mixed parental lines and their cross-pollination so that hybrids could be produced with distinct features (Yadav and Rai, 2013). However, the success rate of the method was limited to 10–20%. Biotechnological techniques and advanced equipment make it possible to create a large number of pearl millet hybrids with desired features. Advancement in technical methods allows the scientists to effect changes as well as molecular modifications at the desired location in the genome. The development of male sterility at cytoplasmic-nuclear level and its release in the form of Tift 18A and Tift 23A opened a new era in the field of hybrid production (Burton, 1983; Yadav and Rai, 2013).

Lytic activity is the feature of lytic factors that are components with small to large molecular weight and have the capability to act on the cellular walls and finally lyse them (Larena and Melgarejo, 1996, 1993). Scientific reports suggest that specific components with lytic activity produced plants as well as microorganisms (Lozovaya et al., 1998; Umesha et al., 2000). Specific components which possess lytic activity are responsible for the occurrence of the exolysis of fungal pathogens (Campbell, 1989; Campbell and Ephgrave, 1993).

Umesha et al. (2000) studied the presence of lytic factors with specific activity in the coleoptiles region of pearl millet cultivars (susceptible and resistant). They found that resistant cultivars of pearl millet possess a high level of lytic factors (100%) as compared to susceptible ones, where the contribution of lytic factors is 20% only. Owing to the presence of components with lytic activity, plants become strong enough to fight against diseases such as downy mildew. Lavanya et al. (2018) studied the effect of lipo-polysaccharide elicitors on downy mildew disease in pearl millet. Their observations indicate that lipo-polysaccharide elicitors protect the pearl millet crop from downy mildew disease by activating pearl millet defense mechanisms. Nitric oxide concentration is a determinant factor for the occurrence of defense mechanisms in pearl millet.

## 3.6 CONCLUSIONS

With respect to climactic changes and soil profile, there is an urgent requirement for a cereal crop that can tolerate harsher conditions and is strong enough to provide nutrient-rich grains. The demand for food products of a natural origin is increasing, as they pose no risk to health and provide nutrients in line with the daily requirements. Pearl millet grain is rich in specific phytochemicals, amino acids and important minerals, which makes it a perfect choice for agriculture systems, food and pharmacological industries. Various biotechnological tools are being applied to pearl millet grain either to improve its nutritional profile or to induce genetic modification for disease resistant features. Owing to the presence of certain specific phytochemicals, whole grain pearl millet or its flour could be used in the preparation of bakery products and other functional food products.

## REFERENCES

Adebiyi, J. A., Obadina, A. O., Adebo, O. A., Kayitesi, E. 2017. Comparison of nutritional quality and sensory acceptability of biscuits obtained from native, fermented, and malted pearl millet (*Pennisetum glaucum*) flour. *Food Chemistry* 232: 210–217.

Alam, M. A., Subhan, N., Hossain, H., Hossain, M., Reza, H. M., Rahman, M. M., Ullah, M. O. 2016. Hydroxycinnamic acid derivatives: A potential class of natural compounds for the management of lipid metabolism and obesity. *Nutrition & Metabolism* 13: 27. DOI: 10.1186/s12986-016-0080-3.

Avasthi, H., Pathak, R. K., Pandey, N., Arora, S., Mishra, A. K., Gupta, V. K., Ramteke, P. W., Kumar, A. 2018. Transcriptome-wide identification of genes involved in Ascorbate-Glutathione cycle (Halliwell–Asada pathway) and related pathway for elucidating its role in antioxidative potential in finger millet (*Eleusine coracana* (L.)). *3 Biotech* 8: 499.

Bei, Q., Chen, G., Lu, F., Wu, S., Wu, Z. 2018. Enzymatic action mechanism of phenolic mobilization in oats (*Avena sativa* L.) during solid-state fermentation with *Monascus anka*. *Food Chemistry* 245: 297–304.

Bento-Silva, A., Koistinen, V. M., Mena, P., Bronze, M. R., Hanhineva, K., Sahlstrom, S., Kitryte, V., Moco, S., Aura, A. 2019. Factors affecting intake, metabolism and health benefits of phenolic acids: Do we understand individual variability? *European Journal of Nutrition* 1–19. DOI: 10.1007/s00394-019-01987-6.

Bhanja, T., Kumari, A., Banerjee, R. 2009. Enrichment of phenolics and free radical scavenging property of wheat koji prepared with two filamentous fungi. *Bioresource Technology* 100: 2861–2866.

Bora, P., Ragaee, S., Marcone, M. 2019. Effect of parboiling on decortication yield of millet grains and phenolic acids and in vitro digestibility of selected millet products. *Food Chemistry* 274: 718–725.

Boz, H. 2015. p-Coumaric acid in cereals: Presence, antioxidant and antimicrobial effects. *International Journal of Food Science and Technology* 50: 2323–2328.

Brul, S., Coote, P. 1999. Preservative agents in foods. Mode of action and microbial resistance mechanisms. *International Journal of Food Microbiology* 50: 1–17.

Burton, G. W. 1983. Breeding pearl millet. *Plant Breeding Reviews* 1: 162–182.

Busso, C. S., Devos, K. M., Ross, G., Mortimore, M., Adams, W. M., Ambrose, M. J., Alldrick, S., Gale, M. D. 2000. Genetic diversity within and among landraces of pearl millet (Pennisetum glaucum) under farmer management in West Africa. *Genetic Resources and Crop Evolution* 47: 561–568.

Campbell, R. 1989. *Biological Control of Microbial Plant Pathogens*. Cambridge, UK: Cambridge University Press.

Campbell, R., Ephgrave, J. M. 1993. Effect of bentonite clay on the growth of *Gaeumannomyces graminis* var. *tritici* and its interaction with antagonistic bacteria. *Journal of General Microbiology* 129: 771–777.

Casagrande, M., Zanela, J., Junior, A. W., Busso, C., Wouk, J., Lurckevicz, G., Montanher, P. F., Yamashita, F., Malfatti, C. R. M. 2018. Influence of time, temperature and solvent on the extraction of bioactive compounds of *Baccharis dracunculifolia*: In vitro antioxidant activity, antimicrobial potential and phenolic compound quantification. *Industrial Crops and Products* 125: 207–219.

Chandrasekara, A., Shahidi, F. 2011. Bioactivities and antiradical properties of millet grains and hulls. *Journal of Agricultural and Food Chemistry* 59: 9563–9571.

Cheok, C. Y., Chin, N. L., Yusof, Y. A., Talib, R. A., Law, C. L. 2012. Optimization of total phenolic content extracted from *Garcinia mangostana* Linn. hull using response surface methodology versus artificial neural network. *Industrial Crops and Products* 40: 247–253.

Dhull, S. B., Kaur, P., Purewal, S. S. 2016. Phytochemical analysis, phenolic compounds, condensed tannin content and antioxidant potential in Marwa (*Origanum majorana*) seed extracts. *Resource Efficient Technologies* 2: 168–174.

Dias-Martins, A. M., Pessanha, K. L. F., Pacheco, S., Rodrigues, J. A. S., Carvalho, C. W. P. 2018. Potential use of pearl millet (*Pennisetum glaucum* (L.) R. Br.) in Brazil: Food security, processing, health benefits and nutritional products. *Food Research International* 109: 175–186.

Freitas, E. R., Fernandes, D. R., Souza, D. H., Dantas, F. D., Santos, R. C., Oliveira, G. B., Cruz, C. E. B., Braz, N. M., Camara, L. F., Nascimento, G. A. 2017. Effect of *syzygium cumini* leaves on laying hens performance and egg quality. *Anais da Academia Brasileira de Ciencias* 89: 2479–2484.

Gale, M. D., Devos, K. M., Zhu, J. H., Allouis, S., Couchman, M. S., Liu, H., Pittaway, T. S., Qi, X. Q., Kolesnikova-Allen, M., Hash, C. T. 2005. New molecular marker technologies for pearl millet improvement. *Journal of SAT Agricultural Research* 1: 1–7.

Gan, R. Y., Chan, C. L., Yang, Q. Q., Li, H. B., Zhang, D., Ge, Y. Y., Gunaratne, A., Ge, J., Corke, H. 2019. Bioactive compounds and beneficial functions of sprouted grains. In (eds.) Hao Feng, Boris Nemzer, Jonathan W. DeVries, *Sprouted Grains*,

*Nutritional Value, Production and Applications,* AACC International: Woodhead Publishing, Elsevier 191–246.

Gandhi, G., Jothi, G., Antony, P. J., Balakrishna, K., Paulraj, M., Ignacimuthu, S., Stalin, A., Al-Dhabi, A. 2014. Gallic acid attenuates high-fat diet fed streptozotocin-induced insulin resistance via partial agonism of PPAR gamma in experimental type 2 diabetic rats and enhances glucose uptake through translocation and activation of GLUT4 in PI3K/p-Akt signaling pathway. *European Journal of Pharmacology* 745: 201–216.

Gani, A., Wani, S. M., Masoodi, F. A., Hameed, G. 2012. Wholegrain cereal bioactive compounds and their health benefits: A review. *Journal of Food Process and Technology* 3: 146.

Godstime, C., Felix, O., Augustina, O., Christopher, O. 2014. Mechanisms of antimicrobial actions of phytochemicals against enteric pathogens. *Journal of Pharmaceutical, Chemical and Biological Sciences* 2: 77–85.

Graf, E. 1992. Antioxidant potential of ferulic acid. *Free Radical Biology and Medicine* 13: 435–448.

Heydaryinia, A., Veissi, M., Sadadi, A. 2011. A comparative study of the effects of the two preservatives, sodium benzoate and potassium sorbate on *Aspergillus niger* and *Penicillium notatum*. *Jundishapur Journal of Microbiology* 4: 301–307.

Iammarino, M., Di-Taranto, A., Palermo, C., Muscarella, M. 2011. Survey of benzoic acid in cheeses: Contribution to the estimation of an admissible maximum limit. *Food Additives & Contaminants: Part B* 4: 231–237.

Irakli, M., Chatzopoulou, P., Ekateriniadou, L. 2018. Optimization of ultrasound-assisted extraction of phenolic compounds: Oleuropein, phenolic acids, phenolic alcohols and flavonoids from olive leaves and evaluation of its antioxidant activities. *Industrial Crops and Products* 124: 382–388.

Isaksson, H., Tillander, I., Andersson, R., Olsson, J., Fredriksson, H., Webb, D. L., Aman, P. 2012. Whole grain rye breakfast-sustained satiety during three weeks of regular consumption. *Physiology & Behavior* 105: 877–884.

Joshi, S. S., Howell, A. B., D'Souza, D. H. 2014. *Cronobacter sakazakii* reduction by blueberry proanthocyanidins. *Food Microbiology* 39: 127–131.

Kaur, P., Purewal, S. S., Sandhu, K. S., Kaur, M. 2019. DNA damage protection: An excellent application of bioactive compounds. *Bioresources and Bioprocessing* 6: 1–11.

Kaur, P., Purewal, S. S., Sandhu, K. S., Kaur, M., Salar, R. K. 2019. Millets: A cereal grain with potent antioxidants and health benefits. *Journal of Food Measurement and Characterization* 13: 793–806.

Kern, S. M., Bennett, R. N., Mellon, F. A. 2003. Absorption of hydroxycinnamates in humans after high-bran cereal consumption. *Journal of Agricultural and Food Chemistry* 51: 6050–6055.

Krebs, H. A., Wiggins, D., Stubbs, M. 1983. Studies on the mechanism of the antifungal action of benzoate. *Biochemical Journal* 214: 657–663.

Kubo, I., Fujita, K., Nihei, K., Masuoka, N. 2003. Non-antibiotic antibacterial activity of dodecyl gallate. *Bioorganic & Medicinal Chemistry* 11: 573–580.

Kumar, A., Tomer, V., Kaur, A., Kumar, V., Gupta, K. 2018. Millets: A solution to agrarian and nutritional challenges. *Agriculture & Food Security* 7: 31.

Kumar, C. T. M., Sabikhi, L., Singh, A. K., Raju, P. N., Kumar, R., Sharma, R. 2019. Effect of incorporation of sodium caseinate, whey protein concentrate and transglutaminase on the properties of depigmented pearl millet based gluten free pasta. *LWT Food Science and Technology* 103: 19–26.

Larena, I., Melgarejo, P. 1996. Biological control of *Monilinia luxa* and *Fusarium oxysporum* f.sp. *lycopersici* by a lytic enzyme producing. *Penicillium purpurogenum, Biological Control* 6: 361–367.

Larena, T., Melgarejo, P. 1993. The lytic enzymatic complex of *Penicillium purpurogenum* and its effects on *Monilinia laxa*. *Mycological Research* 97: 105–110.

Lattimore, P., Walton, J., Bartlett, S., Hackett, A., Stevenson, L. 2010. Regular consumption of a cereal breakfast. Effects on mood and body image satisfaction in adult non-obese women. *Appetite* 55: 512–521.

Li, M. G., Ni, F., Wang, Y. L., Xu, S. D., Zhang, D. D., Chen, S. H., Wang, L. 2009. Sensitive and facile determination of catechol and hydroquinone simultaneously under coexistence of resorcinol with a Zn/Al layered double hydroxide film modified glassy carbon electrode. *Electroanalysis* 21: 1521–1526.

Lin, B. H., Dong, D., Carlson, A., Rahkovsky, I. 2017. Potential dietary outcomes of changing relative prices of healthy and less healthy foods: The case of ready-to-eat breakfast cereals. *Food Policy* 68: 77–88.

Liu, C. J., Witcombe, J. R., Pittaway, T. S., Nash, M., Hash, C. T., Busso, C. S., Gale, M. D. 1994. An RFLP-based genetic map in pearl millet (*Pennisetum glaucum*). *Theoretical and Applied Genetics* 89: 481–487.

Liu, Y. H., Qin, G. W., Fang, J. G., Wu, X. Y. 2002. Screening of chemical constituents with anti endotoxin activity from radix isatidis. *Herbal Medicine* 21: 74–75.

Liyana-Pathirana, C., Shahidi, F. 2005. Optimization of extraction of phenolic compounds from wheat using response surface methodology. *Food Chemistry* 93: 47–56.

Loader, T. B., Taylor, C. G., Zahradka, P., Jones, P. J. H. 2017. Chlorogenic acid from coffee beans: Evaluating the evidence for a blood pressure-regulating health claim. *Nutrition Reviews* 75: 114–133.

Lozovaya, V. V., Waranyuwat, A., Widholmj, M. 1998. β-1,3-glucanases and resistance to *Aspergillus flavus* infection in maize. *Crop Science* 38: 936–943.

Lavanya, S. N., Udayashankar, A. C., Raj, S. N., Mohan, C. D., Gupta, V. K., Tarasatyavati, C., Srivastava, R., Nayaka, S. C. 2018. Lipopolysaccharide-induced priming enhances NO-mediated activation of defense responses in pearl millet challenged with *Sclerospora graminicola*. *3 Biotech* 8: 475 doi:10.1007/s13205-018-1501-y

Malleshi, N. G., Klopfenstein, C. F. 1998. Nutrient composition, amino acid and vitamin contents of malted sorghum, pearl millet, finger millet and their rootlets. *International Journal of Food Sciences and Nutrition* 49: 415–422.

Manuja, R., Sachdeva, S., Jain, A., Chaudhary, J. 2013. A comprehensive review on biological activities of p-hydroxy benzoic acid and its derivatives. *International Journal of Pharmaceutical Sciences Review and Research* 22: 109–115.

Meng, S., Cao, J., Feng, Q., Peng, J., Hu, Y. 2013. Roles of chlorogenic acid on regulating glucose and lipids metabolism: A review. *Evidence-Based Complementary and Alternative Medicine* 2013: 1–11. DOI: 10.1155/2013/80145799.

Nani, A., Belarbi, M., Soualem, Z., Ghanemi, F. Z., Borsali, N., Amamou, F. 2011. Study of the impact of millet (*Pennisetum glaucum*) on the glucidic metabolism in diabetic rats. *Annals of Biological Research* 2: 21–23.

Nesvera, J., Rucka, L., Patek, M. 2015. Chapter Four – Catabolism of phenol and its derivatives in bacteria: Genes, their regulation, and use in the biodegradation of toxic pollutants. *Advances in Advanced Microbiology* 93: 107–160.

Omwamba, M., Hu, Q. 2009. Antioxidant capacity and antioxidative compounds in barley (*Hordeum vulgare* L.) grain optimized using response surface methodology in hot air roasting. *European Food Research and Technology* 229: 907–914.

Ou, S., Kwok, K. C. 2004. Ferulic acid: Pharmaceutical functions, preparation and applications in foods. *Journal of the Science of Food and Agriculture* 84: 1261–1269.

Oussaid, S., Chibane, M., Madani, K., Amrouche, T., Achat, S., Dahmoune, F., Houali, K., Rendueles, M., Diaz, M. 2017. Optimization of the extraction of phenolic compounds from Scirpus holoschoenus using a simplex centroid design for antioxidant and antibacterial applications. *LWT Food Science and Technology* 86: 635–642.

Papatsiros, V. G., Tassis, P. D., Tzika, E. D., Papaioannou, D. S., Petridou, E., Alexopoulos, C., Kyriakis, S. C. 2011. Effect of benzoic acid and combination of benzoic acid with a probiotic containing *Bacillus cereus* var. *toyoi* in weaned pig nutrition. *Polish Journal of Veterinary Sciences* 14: 117–125.

Postemsky, P. D., Bidegain, M. A., Gonzalez-Matute, R., Figlas, N. D., Cubitto, M. A. 2017. Pilot-scale bioconversion of rice and sunflower agro-residues into medicinal mushrooms and laccase enzymes through solid-state fermentation with *Ganoderma lucidum*. *Bioresource Technology* 231: 85–93.

Purewal, S. S., Sandhu, K. S., Salar, R. K., Kaur, P. 2019. Fermented pearl millet: A product with enhanced bioactive compounds and DNA damage protection activity. *Journal of Food Measurement and Characterization* 13: 1479–1488.

Rao, M. V. S. S. T. S., Muralikrishna, G. 2001. Non-starch polysaccharides and bound phenolic acids from native and malted finger millet (Ragi, *Elucine coracana* Indaf-15). *Food Chemistry* 72: 187–192.

Rao, M. V. S. S. T. S., Muralikrishna, G. 2002. Evaluation of the antioxidant properties of free and bound phenolic acids from native and malted finger millet (Ragi, *Eleusine coracana* Indaf-15). *Journal of Agricultural and Food Chemistry* 50: 889–892.

Rosa, N. N., Dufour, C., Lullien-Pellerin, V., Micard, V. 2013. Exposure or release of ferulic acid from wheat aleurone: Impact on its antioxidant capacity. *Food Chemistry* 141: 2355–2362.

Roy, A. J., Prince, P. S. M. 2013. Preventive effects of p-coumaric acid on cardiac hypertrophy and alterations in electrocardiogram, lipids, and lipoproteins in experimentally induced myocardial infarcted rats. *Food and Chemical Toxicology* 60: 348–354.

Ruan, Z. P., Zhang, L. L., Lin, Y. M. 2008. Evaluation of the antioxidant activity of *syzygium cumini* leaves. *Molecules* 13: 2545–2556.

Salar, R. K., Certik, M., Brezova, V. 2012. Modulation of phenolic content and antioxidant activity of maize by solid state fermentation with *Thamnidium elegans* CCF-1456. *Biotechnology and Bioprocess Engineering* 17: 109–116.

Salar, R. K., Purewal, S. S. 2016. Improvement of DNA damage protection antioxidant activity of biotransformed pearl millet (*Pennisetum glaucum*) cultivar PUSA-415 using *Aspergillus oryzae* MTCC 3107. *Biocatalysis and Agricultural Biotechnology* 8: 221–227.

Salar, R. K., Purewal, S. S. 2017. Phenolic content, antioxidant potential and DNA damage protection of pearl millet (*Pennisetum glaucum*) cultivars of North Indian region. *Journal of Food Measurement and Characterization* 11: 126–133.

Salar, R. K., Purewal, S. S., Bhatti, M. S. 2016. Optimization of extraction conditions and enhancement of phenolic content and antioxidant activity of pearl millet fermented with *Aspergillus awamori* MTCC-548. *Resource Efficient Technologies* 2: 148–157.

Salar, R. K., Purewal, S. S., Sandhu, K. S. 2017a. Fermented pearl millet (*Pennisetum glaucum*) with in vitro DNA damage protection activity, bioactive compounds and antioxidant potential. *Food Research International* 100: 204–210.

Salar, R. K., Purewal, S. S., Sandhu, K. S. 2017b. Relationships between DNA damage protection activity, total phenolic content, condensed tannin content and antioxidant potential among Indian barley cultivars. *Biocatalysis and Agricultural Biotechnology* 11: 201–206.

Salar, R. K., Purewal, S. S., Sandhu, K. S. 2017c. Bioactive profile, free-radical scavenging potential, DNA damage protection activity, and mycochemicals in *Aspergillus awamori* (MTCC 548) extracts: A novel report on filamentous fungi. *3 Biotech* 7: 164.

Salar, R. K., Sharma, P., Purewal, S. S. 2015. *In vitro* antioxidant and free radical scavenging activities of stem extract of *Euphorbia trigona* Miller. *TANG [Humanitas Medicine]* 5: 1–6.

Saleh, A. S. M., Zhang, Q., Chen, J., Shen, Q. 2013. Millet Grains: Nutritional quality, processing, and potential health benefits. *Comprehensive Reviews in Food Science and Food Safety* 12: 281–295.

Sandhu, K. S., Siroha, A. K. 2017. Relationships between physicochemical, thermal, rheological and in vitro digestibility properties of starches from pearl millet cultivars. *LWT-Food Science and Technology* 83: 213–224.

Shaikh, M., Haider, S., Ali, T. M., Husnain, A. 2019. Physical, thermal, mechanical and barrier properties of pearl millet starch films as affected by levels of acetylation and hydroxypropylation. *International Journal of Biological Macromolecule* 124: 209–219.

Sharma, P., Gujral, H. S. 2010. Antioxidant and polyphenol oxidase activity of germinated barley and its milling fractions. *Food Chemistry* 120: 673–678.

Shin, H. Y., Kim, S. M., Lee, J. H., Lim, S. T. 2019. Solid-state fermentation of black rice bran with *Aspergillus awamori* and *Aspergillus oryzae*: Effects on phenolic acid composition and antioxidant activity of bran extracts. *Food Chemistry* 272: 235–241.

Singh, R. B., Khan, S., Chauhan, A. K., Singh, M., Jaglan, P., Yadav, P., Takahashi, T., Juneja, L. R. 2019. Millets as functional food, a gift from Asia to Western World. *The Role of Functional Food Security in Global Health*, 457–468. DOI: 10.1016/B978-0-12-813148-0.00027-X.

Singh, S., Kaur, M., Sogi, D. S., Purewal, S. S. 2019. A comparative study of phytochemicals, antioxidant potential and in vitro DNA damage protection activity of different oat (*Avena sativa*) cultivars from India. *Journal of Food Measurement and Characterization* 13: 347–356.

Siroha, A. K., Sandhu, K. S., Kaur, M. 2016. Physicochemical, functional and antioxidant properties of flour from pearl millet varieties grown in India. *Journal of Food Measurement and Characterization* 10: 311–318.

Solecka, D. 1997. Role of phenylpropanoid compounds in plant responses to different stress factors. *Acta Physiologiae Plantarum* 19: 257–268.

Tesaki, S., Tanabe, S., Ono, H., Fukushi, E., Kawabata, J., Watanabe, M. 1998. 4-hydroxy-3-nitrophenylacetic and sinapic acids as antibacterial compounds from mustard seeds. *Bioscience, Biotechnology and Biochemistry* 62: 998–1000.

Umesha, S., Shylaja, M., Dharmesh, H., Shetty, S. 2000. Lytic activity in pearl millet: Its role in downy mildew disease Resistance. *Plant Science* 157: 33–41.

Ververidis, F., Trantas, E., Douglas, C., Vollmer, G., Kretzschmar, G., Panopoulos, N. 2007. Biotechnology of flavonoids and other phenylpropanoid-derived natural products. Part I: Chemical diversity, impacts on plant biology and human health. *Biotechnology Journal* 2: 1214–1234.

Wang, Y., Compaore-Sereme, D., Sawadogo-Lingani, H., Coda, R., Katina, K., Maina, N. H. 2019. Influence of dextran synthesized in situ on the rheological, technological and nutritional properties of whole grain pearl millet bread. *Food Chemistry* 285: 221–230.

Xiong, Y., Zhang, P., Luo, J., Johnson, S., Fang, Z. 2019. Effect of processing on the phenolic contents, antioxidant activity and volatile compounds of sorghum grain tea. *Journal of Cereal Science* 85: 6–14.

Yadav, O. P., Rai, K. N. 2013. Genetic improvement of pearl millet in India. *Agricultural Research* 2: 275–292.

Yoon, S., Kang, S., Shin, H., Kang, S., Kim, J., Ko, H., Kim, S. 2013. p-Coumaric acid modulates glucose and lipid metabolism via AMP-activated protein kinase in L6 skeletal muscle cells. *Biochemical and Biophysical Research Communications* 432: 553–557.

Zhang, F., Wan, X., Zheng, Y., Sun, L., Chen, Q., Zhu, X., Guo, Y., Liu, M. 2013. Effects of nitrogen on the activity of antioxidant enzymes and gene expression in leaves of Populus plants subjected to cadmium stress. *Journal of Plant Interactions* 9: 599–609.

# 4 Effects of Different Milling Processes on Pearl Millet

*Sneh Punia, Anil Kumar Siroha and Sanju Bala Dhull*

## CONTENTS

## 4.1 INTRODUCTION

Pear millet, being a good source of complex carbohydrates, dietary fibers and phytochemicals, is known as nutricereal. However, because of non-availability of ready to use and eat food products from pearl millet, it is only used as food for poor and traditional consumers. Generally, pearl millet is used to prepare porridge and roti either from coarse grain or finely milled grain. However, its nutrient availability is constrained by antinutritional factors such as polyphenols and phytic acid. Phytic acid binds with minerals (iron, calcium, magnesium and zinc), proteins and starch and reducing the absorption compounds in human consumption results in reduced bioavailability (El-Hag et al., 2002), whereas polyphenols interfere with the intestinal absorption of minerals (Abdelrahman et al., 2005). Hence, it is important to reduce the phytic acid and polyphenol content to make the nutritional benefits of this grain available. Also, pearl millet

51

may be stored for longer periods without any alteration in quality if the kernel remains intact (Kachare and Chavan, 1992) but once the pearl grain is dehulled and ground, the quality of grain starts to deteriorate and the flour becomes rancid. The lipidic profile may change due to formation of free fatty acids and peroxides, and hydrolytic and oxidative changes in flour causes bitterness.

To overcome these problems, the pearl millet grains are subjected to pre-milling processing to improve sensorial and edible qualities (Liu et al., 2012). Several studies on millets have reported that different processing may reduce or increase the nutrients, depending on the severity of heat treatment, time of exposure and the type of cereal tested (Hegde and Chandra, 2005; Towo et al., 2003; Zielinski et al., 2006). Various traditional household food processing and preparation methods can be used to enhance the bioavailability of micronutrients in plant-based diets, such as dehulling, soaking, milling, cooking, germination, fermentation and several thermal processes.

## 4.2  IMPORTANCE OF PROCESSING

Coarse fibrous grains and poor shelf life of pearl millet flour are major constraints to the commercialization of this crop. Colored pigments and the characteristic astringent flavor in pearl millet are additional constraints to commercialization (Desikachar, 1975). When used for preparing shelf-stable food products, the grain requires a processing technology that can remove the germ with little loss in the grain. Pre-milling treatment of pearl millet grain improves nutritive quality, reduces antinutritional factors, and increases digestibility and consumer acceptability (Table 4.1). Optimal heat treatment that destroys lipolytic enzymes without affecting the natural protective oxidant presence has the potential to extend the shelf life of the flour and food products. During the milling process, bran, which has lower ash and silica contents, is removed

**TABLE 4.1**

**Importance of millet processing**

| Factors | Importance | References |
|---|---|---|
| Nutritonal bioavailability | Improves thiamine, niacin, total lysine, protein fractions, sugars, soluble dietary fiber, and *in vitro* availability of Ca, Fe, and Zn of food blends | Arora et al. (2011) |
| Shelf life | Inactivates lipase activity | Yadav et al. (2012) |
| Organoleptic properties | Processing optimizes the appearance, taste and texture of foods to meet the needs of consumers | Sandhu et al. (2016) |
| Antinutritional factors | Reducing antinutritional compounds like phytic acid, tannins, and polyphenols, which form complexes with protein | Hassan et al. (2006) |
| Digestibility | Improved digestibility and maximizes bioavailability of iron and zinc from cereals and pulses | Lestienne, Besanon et al. (2005) |

**TABLE 4.2**

**Different methods for processing**

| Methods | Definition | References |
|---|---|---|
| Decortication | A technique of removing pericarp from grains of cereals | Scheuring and Rooney (1979) |
| Blanching | A hydrothermal treatment which induces significant changes in chemical composition, affecting the bioacessibility and the concentration of nutrients and health-promoting compounds | Pellegrini et al. (2010) |
| Acid treatment | Treats the decorticated seed with an acetic, fumaric, or tartaric and also with the extracts of natural tamarind | Hadimani and Malleshi (1993) |
| Germination | Induces the synthesis or activation of a range of hydrolytic enzymes in the germinated grain, resulting in structural modification or synthesis of new compounds with high bioactivity or nutritional value | Wang et al. (2014) |
| Fermentation | A process in which plants and animal tissues are subjected to the action of microbes and enzymes to give desirable changes and to modify food quality | Sandhu, Punia et al. (2017) |
| Microwacve cooking | Microwave cooking has gained considerable importance as an energy-saving, convenient and time-saving cooking method in which heat is generated from within the food through a series of molecular vibrations | Cross et al. (1982) |
| Extrusion cooking | A rapid processing method involving high temperature and pressure and short time, and is used to prepare a variety of processed foods | Sharma et al. (2012) |
| Roasting | A high temperature short time (HTST) heat treatment process and the temperatures used are normally in the range of 280–350°C | Murthy et al. (2008) |
| Toasting | A rapid processing method that uses dry heat for short periods of time | Sandhu, Godara et al. (2017) |
| Milling | An intermediate step in post-production of grain and defined as an act or process of grinding the grain into flour | Bender (2006) |

and it may be efficiently used as a source of dietary fiber. Bran contains a high proportion of soluble dietary fiber and could be tapped for hypocholesterolemic and hypoglycemic effects. By applying good milling technologies to produce suitable flours, meals, and grit, a wide range of baked and steamed food products may be prepared. Different methods of processing and their effects on nutritional quality are described in Tables 4.2 and 4.3.

## 4.3 STEPS INVOLVED IN MILLING

### 4.3.1 DEHULLING/DECORTICATION

The tough outer seed coat and its associated characteristic flavor (Malleshi et al., 1986), cultural tradition and non-availability of processed millet

**TABLE 4.3**

**Effect of different treatments on nutritional quality of pearl millet**

| Methods | Effects on nutritional quality | References |
|---|---|---|
| Decortication | Reduces phytates (4% and 8%) and substantial losses of zinc, iron, fibre and iron binding phenolics | Lestienne et al. (2007) |
| Soaking | Reduces phytic acid concentration | Duhan et al. (1989) |
| | leaching of iron and zinc ions | Saharan et al. (2001) |
| | Improves the grain color and softens the grain, resulting in easier milling | (Taylor, 2004) |
| | Reduces phytate and zinc content | Lestienne, Besanon et al. (2005) |
| | Reduces fat, phytic acid, lessens gelation capacity and viscosity | Ocheme and Chinma (2008) |
| | Reduces phytate and flavonoids content up to 81% and 63%, | Jha et al. (2015) |
| | phytic acid content decreased significantly ($p < 0.05$) from its initial value of 207.23 to 198.96 mg and 183.49 mg after 9 and 12 h of soaking | Sihag et al. (2015) |
| Cooking | Decreases the β-carotene, iron and ash contents | Kataria et al. (1989) and Sihag et al. (2015) |
| | Decreases polyphenols, phytate and increases protein digestibility | Pawar and Machewad (2006) |
| | Reduces polyphenols up to 41.66% | Hithamani and Srinivasan (2014) |
| Acid treatment | Improves the product quality by reducing polyphenols and other antinutritional factors, thereby also increasing consumer acceptability | Hadimani and Malleshi (1993) |
| Germination | Decreases the level of tannins from 1.6% to 0.83% | Opoku et al. (1981) |
| | Decrease in the antinutritional factors | Gupta and Sehgal (1991). |
| | Significant reduction of polyphenols and phytic acid | Archana et al. (1998) |
| | Increased mineral bioavailability | Pawar and Parlikar (1990); Badau et al. (2005); Arora et al. (2003) |
| | Softening the kernel structure, improving its nutritional value, and reducing antinutritional effects | Tian et al. (2010) |
| | An increase of 13.2 and 16.3% in protein content of pearl millet. | Hassan et al. (2006) |
| | Iron and calcium content significantly increased | Sushma et al. (2008). |
| | Improvement in the contents of thiamine, niacin, total lysine, protein fractions, sugars, soluble dietary fibre and in vitro availability of Ca, Fe and Zn | Arora et al. (2011). |
| | Reduces antinutrients, thereby improving nutritional and functional properties of pearl millet | Suma and Urooj (2014) |
| | Bioavailability of nutrients improved | Singh et al. (2015). |

*(Continued)*

**TABLE 4.3 (Cont.)**

| Methods | Effects on nutritional quality | References |
|---|---|---|
| Fermentation | Increases pepsin digestibility of millet protein, decreases the concentration of phytic acid and polyphenols | Mahajan and Chauhan (1987) |
| | Improves the availability of minerals | Khetarpaul and Chauhan (1989) |
| | Reduction in protein content | Abdalla et al. (1996) |
| | A significant decrease in polyphenols from 352.64 to 278.72 mg/100 g after 36 h | Tiwari et al. (2014) |
| Milling | Improves the availability of antioxidant compound and accessibility of digestion enzymes to components that are bound with food matrix | (Liukkonen et al., 2003; Nagah and Seal, 2005; Parada and Aguilera, 2007; Prom-u-thai et al., 2006). |

products similar to rice or wheat are the main reasons for a lack of popularity of millet foods among rice and wheat eaters (Malleshi and Hadimani, 1993). Dehulling, also known as decortication/pearling, is a technique used to remove the pericarp from cereal grains (Figure 4.1). It improves texture, color, nutritional quality, and cooking quality of respective grains (Scheuring and Rooney, 1979). Both whole grains and dehulled grains of pearl millet are used for preparing various types of food products. Grains of pearl millet are globular in shape and have corny endosperm and thick pericarp that are relatively easy to decorticate to the extent of 12% to 30% (Rai et al., 2008). As reported by Rai et al. (2008), decortication beyond 30% causes substantial loss of ash, fat, micronutrients, fiber, proteins and amino acids. However, phytates are present in the outer fraction of pearl millet grains, and they are removed during decortication. The fractions of husk in pearl millet and small millet varied from 1.5% to 29.3% (Hadimani and Malleshi, 1993). In earlier times, millets were decorticated by hand pounding, but nowadays, rice milling machines (Singh and Raghuvanshi, 2012) and rice hullers with polishers (Agu et al., 2007) are commonly used for this purpose. An abrasive mill (Ayo and olawale, 2003) and centrifugal sheller (Jaybhaye et al., 2014) may be used to dehull/decorticate small millets. A dehuller (Figure 4.2) designed basically for sorghum but with some modifications for pearl millet, has been manufactured by the Rural Industries Innovation Center (RIIC), Kanye, Botswana, which has a capacity of 400–600 kg/h (Rohrbach and Obilana, 2004).

Serna-Saldivar et al. (1994) reported that although decortication improved protein and dry matter digestibility, due to the removal of the pericarp and germ, reduction in protein, fat, insoluble dietary fiber, ash, lysine and tryptophan was also observed. According to Lestienne et al. (2007), at decortication of 12%, up to 8% of phytates of pearl millet grains were removed from Gampela and IKMP-5 cultivars of pearl millet, respectively. They estimated the losses of nutrients and antinutrients at an 88% extraction rate through abrasive decortication. There were substantial losses of zinc, iron, fiber and iron

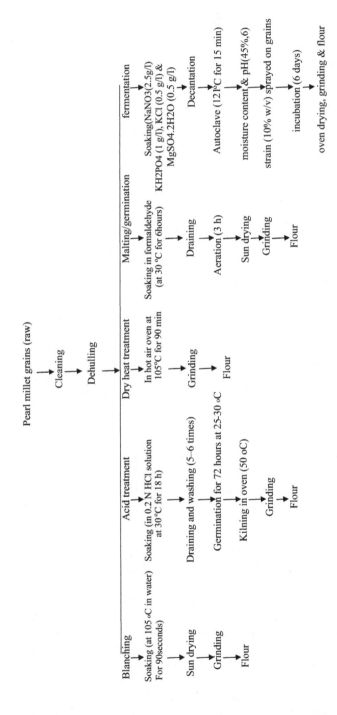

**FIGURE 4.1**  Flow diagram of premilling treatments of pearl millet grains (Bhati et al., 2016; Sandhu et al., 2016)

**FIGURE 4.2** Pearl millet grain dehuller

binding phenolics. As reported by Krishnan et al. (2012), decortication decreases the total mineral content, but increases the bioaccessibility of calcium, iron and zinc by 15, 26 and 24 g/100 g, respectively. Jaybhaye et al. (2014) reported that dehulling coupled with hydrothermal treatment affects the phenolic content and antioxidant potential of millet grains.

### 4.3.2 Soaking

Soaking of grains in plain water is a popular household food processing technique and a common practice to soften texture and hasten the cooking process (Silva et al., 1981). It is used for reducing antinutritional compounds such as phytic acid and phytase activity to improve bioavailability of minerals (Lestienne, Icard-Vernière et al., 2005). Soaking of millet grains can be used as a pre-treatment under optimized conditions to reduce the antinutrition content in the grains and enhance nutrient bioavailability and nutritional quality of millet food products (Saleh et al., 2013). Pearl millet is often soaked in water at ambient temperature during food processing (Mallet and du Plessis, 2001). It improves the grain color and softens the grain, which results in easier milling to produce flour (Taylor, 2004). A reduction in phytic acid concentration during pearl millet soaking was reported by Duhan et al. (1989). Eyzaguirre et al. (2006) concluded that soaking of grains results in a 25% loss of iron, but also facilitates endogenous phytate degradation, particularly when combined with milling and cooking. A study conducted by Lestienne, Mouquet-Rivier et al. (2005) showed that soaking resulted in reduction of phytate and zinc content in pearl millet. This reduction may be attributed to the leaching of iron and zinc ions into the soaking medium (Saharan et al., 2001). Mohamed et al.

(2007) studied the effect of soaking on two pearl millet cultivars (Gazira and Gadarif), and reported reduced phytic acid content from 987.19 to 597.50 mg/100 g for the Gazira cultivar and from 952.51 to 722.20 mg/100 g for the Gadarif cultivar. Ocheme and Chinma (2008) observed that fat, phytic acid, gelation capacity and viscosity of pearl millet decreased significantly (p < 0.05) as a result of soaking and germination. The decreased value of fat in flour samples as a result of soaking and germination may be due to the action of lipolytic enzymes that utilized the fats present. Jha et al. (2015) reported that an acidic medium reduced phytate and flavonoids content up to 81% and 63%, but soaking in an alkaline medium did not cause a substantial effect on zinc content while resulting in decreasing flavonoids by 60% in bran and endosperm-rich fractions of pearl millet. Sihag et al. (2015) reported phytic acid content decreased significantly (p <0.05) from its initial value of 207.23 mg to 198.96 mg and 183.49 mg after 9 and 12 h of soaking, respectively.

### 4.3.3 PROCESSING

### 4.3.3.1 Blanching

Blanching/cooking is employed to reduce the level of antinutrients such as phytic acid and polyphenols (Sihag et al., 2015). Hithamani and Srinivasan (2014) investigated the effect of domestic processing on the polyphenol content in pearl millet and observed that sprouting and pressure-cooking reduced polyphenols by 33.52% and 41.66%, respectively. Sihag et al. (2015) reported that cooking led to a significant (p < 0.05) decrease in the β-carotene, iron, and ash contents of the pearl millet just after 2 min of cooking. Another reason attributed to the reduction of these compounds is their thermal degradation during pressure-cooking (Kataria et al., 1989; Kataria et al. 1992). It is found that a combination of different processes such as dehulling, soaking and cooking significantly reduces the amount of antinutrients such as polyphenols and phytate and increases the protein digestibility *in vitro* while also improving the bioavailability of minerals such as iron and zinc (Pawar and Machewad, 2006). Dharmaraj et al. (2014) observed that a reduction in cooking time might be associated with various factors such as smaller size, removal of the seed coat, and a larger surface area, as well as the presence of pre-gelatinized starch in decorticated millet.

### 4.3.3.2 Acid Treatment

The color of pearl millet is an adverse property of the grain for utilization in the preparation of food products. The dark gray color of pearl millet after decortication is bleached by using acids (acetic acid, tartaric acid and fumaric acid) or with natural acidic material such as tamarind (Rai et al., 2008). Hadimani and Malleshi (1993) concluded that acid treatments have the potential to improve the product quality by reducing polyphenolic compounds and many antinutritional compounds, ultimately increasing consumer acceptability. The effects of acid treatment on nutritional and antinutritional components have been studied by various researchers for many years. As reported by Naikare et al. (1986), dilute hydrochloric acid (HCl) is found to be more effective when compared with organic acids such as acetic and citric acid. Arora et al. (2003) concluded that acid treatment carried out by soaking the grains in HCl solution (0 0.2 N) for 6, 12, 18 and 24 h, followed by washing, blanching (98°C for 30 sec) and sun drying (two days) showed significant improvement in the extractability of phosphorus, calcium and iron as the period of acid soaking was prolonged. This improvement of HCl extractability was further accompanied by higher bioavailability of minerals. Bhati et al. (2016) reported that pearl millet grains, when subjected to acid treatment (2, 12, 18, 24 h), revealed lower polyphenols and free fatty acid profiles. Moreover, 18 h acid treatment was found to be most effective in maximum enhancement of *in vitro* iron availability as compared to raw grains (2.19–3.01 mg/100 g). However, lower iron content observed in this study could be associated with leaching of minerals that were present mainly in the pericarp of the seed. Therefore, for the production of pearl millet-based food products having the advantages of a lower antinutritional profile, higher bioavailability of minerals and improved color characteristics, it is necessary to promote the large scale use of acid treatment.

### 4.3.3.3 Malting/Germination

Additionally, germination has been claimed to improve the nutritional quality of cereals and has been used for centuries for the purpose of softening the kernel structure, improving its nutritional value, and reducing antinutritional effects (Tian et al., 2010). Germination is a common household technique carried out at low cost without the use of any sophisticated and expensive equipment. Opoku et al. (1981) showed that germination of pearl millet decreased the level of tannins from 1.6% to 0.83%. It was observed that the decrease in the antinutritional factors of cereal grains was a result of soaking and germination (Gupta and Sehgal, 1991). Besides lowering of antinutrients, germination for 72 h significantly increased the HCl extractability of minerals, which represents mineral bioavailability (Arora et al., 2003; Badau et al., 2005; Pawar and Parlikar, 1990). According to a study conducted by Archana et al. (1998), pearl millet grains, when subjected to malting, showed a significant reduction of polyphenols and phytic acid. This reduction of polyphenols after malting may be associated with the presence of polyphenol oxidase or may arise from the hydrolysis of tannin protein and tannin enzyme complexes, which promotes the elimination of tannin or polyphenols. Hassan et al. (2006) reported an increase of 13.2% and 16.3% in the protein content of pearl millet. Upon germination, the iron and calcium content of pearl millet was also increased, as reported by Sushma et al. (2008). It reduces antinutrients, thereby improving the nutritional and functional properties of pearl millet (Suma and Urooj, 2014). Germination results in biochemical modifications and produce malt with improved nutritional quality that may be used in various traditional recipes

(Saleh et al., 2013). The germination process was found to be responsible for promoting enzymatic activity of sprouted seeds, and, thereby, causing the disintegration of carbohydrates, proteins and lipids into simpler forms. The bioavailability of nutrients also improved significantly as a result of degradation of proteins by protease enzymes (Singh et al., 2015). This processing appreciably improved the *in vitro* protein (14–26%) and starch (86–112%) digestibility in pearl millet (Archana and Kawatra, 2001). Furthermore, the relative *in vitro* solubility of iron was doubled through the germination of pearl millet grains (Eyzaguirre et al., 2006). Suma and Urooj (2014) reported that germination resulted in a significant reduction of phytate and oxalate content, which led to an increase in bioaccessible iron and calcium content. The process also resulted in lowering of antinutrients/mineral molar ratios, which had a positive impact on the bioaccessible mineral content. Germination and probiotic fermentation caused significant improvement in the contents of thiamine, niacin, total lysine, protein fractions, sugars, soluble dietary fiber and *in vitro* availability of calcium (Ca), iron (Fe) and Zinc (Zn) (Arora et al., 2011).

### 4.3.3.4 Thermal Treatment

Lipase activity is the major cause of spoilage of pearl millet meal, so its deactivation before milling improves the meal quality. The application of dry heat to the meal effectively retards lipase activity and minimizes lipid decomposition during storage (Rai et al., 2008). Thermal treatments may reduce or increase the nutritional quality of cereals, depending on heat treatment, time period and the type of cereal tested (Hegde and Chandra, 2005; Zielinski et al., 2006). The effect of heat treatments on the nutritional quality of pearl millet is presented in Table 4.4. Heat energy application is a treatment that has been widely investigated with the aim of deactivating lipase in pearl millet. Pearl millet flour, when stored, develops bitterness and rancidity, which limits its shelf life. These problems occur due to the activity of lipase enzymes which cause the breakdown of glycerides and subsequent increase of the free fatty acid profile (Arora et al., 2002). Various treatments were applied to pearl millet to prevent the development of rancidity in the stored flour: application of dry heat treatment to grain (Chavan and Kachare, 1994) or flour (Kapoor and Kapoor, 1990), and application of thermal treatment for producing shelf stability (Nantanga et al., 2008), which increased the storage life of pearl millet flour up to six days without undue deterioration in quality in ambient conditions (Tiwari et al., 2014). Therefore, the deactivation of lipase activity before milling is necessary for the enhancement of the quality of the meal. This can be achieved by exposing the pearl millet grain to a dry heat treatment, which significantly retards the lipase activity and minimizes the lipid decomposition during storage (Rai et al., 2008). According to Tiwari et al. (2014) pearl millet flour, when subjected to heat treatment carried out at 110°C for 60 sec, showed a significant reduction in fat acidity and the free fatty acids profile. Dry heat treatment of pearl millet grains (100°C for 120 min) showed higher *in vitro* iron availability (2.19–3.58 mg/100 g), along with

## TABLE 4.4
### Effect of heat treatments on nutritional quality of pearl millet

| Thermal treatments | Effects | References |
|---|---|---|
| Dry heat treatment | Reduces bio-availability of minerals and digestibility of protein | Reyden and Selvendran, 1993; Mohamed et al. (2010) |
| | Retards lipase activity and minimizes lipid decomposition during storage | Rai et al. (2008) |
| | Higher *in vitro* iron availability along with maximum retardation of polyphenol | Bhati et al. (2016) |
| Microwave cooking | Decreases lipase activity | Yadav et al. (2012) |
| | Reduces the total polyphenol and flavonoid content | Hithamani and Srinivasan (2014) |
| Extrusion cooking | Reduces the anti-nutritional factors, renders the product microbially safe and enhances consumer acceptability | Nibedita and Sukumar (2003) |
| | Inactivates lipases and enhances the shelf life | Sumathi et al. (2007). |
| | Reduces phytic acid by 43.68%, and polyphenols and increases storage life of pearl millet flour up to 6 days without undue deterioration in quality | Tiwari et al. (2014) |
| Roasting | Reduces phytic acid content of product | Jalgaonkar et al. (2016) |
| | Reduces lysine, methionine, potassium, phosphorus, calcium, magnesium and phenolic contents and improves the functional properties of pearl millet flours | Obadina et al. (2016) |
| Toasting | Results in increase of antioxidant properties | Siroha and Sandhu (2017) |

maximum retardation of polyphenol (675.33–477.93 mg/100 g) and free fatty acid content (44.56–16.23 mg/100 g), than raw grain. However, loss of minerals (phosphorus, iron, calcium) observed during the processing may be attributed to their destruction as a result of heating (Bhati et al., 2016).

Phytic acid in cereals chelates mineral cations and interacts with proteins, forming insoluble complexes, which leads to reduced bioavailability of minerals and reduced digestibility of protein (Mohamed et al., 2010; Reyden and Selvendran, 1993). Jalgaonkar et al. (2016) reported a reduction in phytic acid content of pearl millet in roasted flour and in hydro-thermally treated flour. In a study carried out by Obadina et al. (2016), it was reported that roasting reduced lysine, methionine, potassium, phosphorus, calcium, magnesium and phenolic content of pearl millet flour and improved its functional properties, which might be due to the changes in the bioavailability of nutrients such as proteins in the flour after roasting (Gabrelibanos et al., 2013). Toasting (mild heat treatment) has numerous health benefits and improves product quality (Sandhu, Godara et al., 2017).

Siroha and Sandhu (2017) reported that toasting resulted in an increase in the antioxidant properties of pearl millet. Yadav et al. (2012) reported a significant decrease in lipase activity in pearl millet (p <0.05) with an increase in the duration of microwave exposure. Hithamani and Srinivasan (2014) concluded that microwave heating reduced the total polyphenol and also total flavonoid content in pearl millet. But, on the other hand, the bioaccessibility of identified phenolic compounds increased in the case of pearl millet.

Extrusion cooking is a high temperature short time cooking process, which could be used for processing of starchy and proteinaceous materials (Jaybhaye et al., 2014). This processing also reduces the antinutritional factors, renders the product microbially safe and enhances consumer acceptability (Nibedita and Sukumar, 2003). Besides, in the case of pearl millet, extrusion cooking offers additional benefits; namely, deactivation of lipases and enhancing the shelf life of its products (Sumathi et al., 2007). Tiwari et al. (2014) observed that extrusion of pearl millet significantly reduced phytic acid, by 43.68%, and polyphenol, but had no significant reduction in iron and zinc. The apparent decrease in phytate content during thermal processing may be partly due to the formation of insoluble complexes between phytate and other components, such as phytate protein and phytate–protein–mineral complexes, or to the inositol hexaphosphate hydrolyzed to pentakis and tetraphosphate (Siddhuraju and Becker, 2001).

### 4.3.3.5 Fermentation

Throughout history, fermentation has been used to improve product properties. During fermentation, the grain constituents are modified by the action of both endogenous and bacterial enzymes, thereby affecting their structure, bioactivity and bioavailability (Hole et al., 2012). Fermentation also enhances the levels of bioactive compounds, which can be used to improve product properties (Sandhu and Punia, 2017; Sandhu et al., 2016). Fermentation has been found to increase pepsin digestibility of millet protein, decrease the concentration of phytic acid and polyphenols (Mahajan and Chauhan, 1987) and improve the availability of minerals (Khetarpaul and Chauhan, 1989). Fermentation may have also played a role in an increase in protein content, as it has been observed by other researchers (Alhag, 1999; Obizoba and Atii, 1994). However, Abdalla et al. (1996) observed a non-significant reduction in the protein content of pearl millet during fermentation. Mahajan and Chauhan (1987) reported that endogenous phytase of pearl millet contributed significantly to the reduction of the phytate content of fermented pearl millet flour. Tiwari et al. (2014) found that fermentation caused a significant decrease in polyphenols (352.64–278.72 mg/100 g after 36 h). This result agrees with Dhankher and Chauhan (1987) and Elyas et al. (2002). They reported a decrease in pearl millet polyphenols with increasing fermentation time. Reduction in polyphenols may be due to the activation of polyphenol oxidase (Dhankher and Chauhan, 1987). Salar et al. (2016) envisaged a two-stage optimization of conditions using RSM for the extraction of total phenolic compounds from pearl millet *koji* prepared with *Aspergillus awamori* and reported an enhancement in bioactive compounds. Thus, bio-transformed fermented pearl millet may be used in the preparation of functional foods and novel nutraceuticals for health promotion (Salar et al., 2017).

### 4.3.4 MILLING

Milling is an important and intermediate step in the post-production processing of grain and is defined as an act or process of grinding, especially grinding grain into flour or meal (Bender, 2006). Millet grains are usually milled by an unmotorized grain mill that is cranked by hand or another non-electric method, especially in rural areas for household uses. However, a manual grain mill that has been attached to a gas or electric motor with a pulley system can also be used (Saleh et al., 2013). In pearl millet, a combination of, first, abrasive decortication to remove the bran layers, and then hammer-milling of endosperm is usually used (Kebakile et al., 2007). Nutrients and phytonutrients are not evenly distributed throughout the grain; most of the concentration of nutrients is higher in the outer part of the grain, so differential milling or refining results in reduced nutrient content except for starch (Slavin et al., 1999). The major compositional difference between whole grains and their milled form is a reduction of all nutrients that are stored in the external layer, dietary fiber, and the components associated with fibers, including phytic acid, tannin, polyphenol and some enzyme inhibitors, such as trypsin inhibitors, as well as minerals and some vitamins (García-Estepa et al., 1999). Milling and heat treatment during the making of *chapati* (an unleavened bread) lowered polyphenols and phytic acid and improved the protein and starch digestibility to a significant extent (Chowdhury and Punia, 1997). Milling and refining can improve the availability of antioxidant compounds and their activity because milling breaks the cell wall and grain matrix and improves accessibility of digestion enzymes to components that are bound with the food matrix (Liukkonen et al., 2003; Nagah and Seal, 2005; Parada and Aguilera, 2007; Prom-u-thai et al., 2006).

## 4.4 CONCLUSIONS

The health benefits and nutrition provided by millets is equivalent to other major cereals such as rice, wheat and maize and the technologies used to process it can further improve its quality in terms of nutrition and other edible properties for household consumption. Generally, processing alters the grain quality, improves the nutritional availability and storage stability of the flour, as well as the products themselves, and retains all parts of the whole grains that are beneficial for health and pleasant to consume. Therefore, for encouraging the commercial utilization of pearl millet grains in food formulations and to achieve better food security, it is necessary to use appropriate processing methods. However, this study of millets demands further extensive research in order to achieve improved millet food products that not only provide good health and other beneficial effects, but also taste good, have an extended shelf life and appealing color, and are also economically feasible for all levels of a population.

## REFERENCES

Abdalla, A. A. 1996. The effect of traditional processing on the nutritive value of pearl millet (Doctoral dissertation, M. Sc. Thesis, Faculty of Agriculture: University of Khartoum, Sudan).

Abdelrahman, S. M., El-Maki, H. B., Idris, W. H., Babikar, E. E., and El-Tinay, A. H. 2005. Antinutritional factors content and minerals availability of pearl millet (*Pennisetum glaucum*) as influenced by domestic processing methods and cultivar. *Journal of Food Technology* 3:397–403.

Agu, H. O., Jideani, I. A., and Yusuf, I. Z. 2007. Nutrient and sensory properties of dambu produced from different cereal grains. *Nutrition & Food Science* 37:272–281.

Alhag, M. A. 1999. Effect of fermentation and dehulling on nutritive value and the in vitro protein digestibility of pearl millet. M.Sc. Thesis, Faculty of Agriculture, University of Khartoum, Sudan.

Archana, S. S., and Kawatra, A. 2001. In vitro protein and starch digestibility of pearl millet (*Pennisetum glaucum* L.) as affected by processing techniques. *Nahrung/Food* 45(1):25–27.

Archana, S. S., Sehgal, S., and Kawatra, A. 1998. Reduction of polyphenols and phytic acid content of pearl millet grains by malting and blanching. *Plant Foods for Human Nutrition* 53:93–98.

Arora, P., Sehgal, S., and Kawatra, A. 2002. The role of dry heat treatment in improving the shelf life of pearl millet flour. *Nutrition and Health* 16:331–336.

Arora, P., Sehgal, S., and Kawatra, A. 2003. Content and HCl-extractability of minerals as affected by acid treatment of pearl millet. *Food Chemistry* 80(1):141–144.

Arora, S., Jood, S., and Khetarpaul, N. 2011. Effect of germination and probiotic fermentation on nutrient profile of pearl millet based food blends. *British Food Journal* 113(4):470–481.

Ayo, J. A., and Olawale, O. 2003. Effect of defatted groundnut concentrate on the physico-chemical and sensory quality of fura. *Nutrition & Food Science* 33 (4):173–176.

Badau, M. H., Nkama, I., and Jideani, I. A. 2005. Phytic acid content and hydrochloric acid extractability of minerals in pearl millet as affected by germination time and cultivar. *Food Chemistry* 92:425–435.

Bender, D. A. 2006. *Bender's dictionary of nutrition and food technology*. 8th ed. Abingdon, UK: Woodhead Publishing & CRC Press.

Bhati, D., Bhatnagar, V., and Acharya, V. 2016. Effect of pre-milling processing techniques on pearl millet grains with special reference to in-vitro iron availability. *Asian Journal of Dairy and Food Research* 35:76–80.

Chavan, J. K., and Kachare, D. P. 1994. Effects of seed treatment on lipolytic deterioration of pearl millet flour during storage. *Journal of Food Science and Technology* 31:80–81.

Chowdhury, S., and Punia, D. 1997. Nutrient and antinutrient composition of pearl millet grains as affected by milling and baking. *Food/Nahrung* 41(2):105–107.

Cross, G. A., Fung, D. Y., and Decareau, R. V. 1982. The effect of microwaves on nutrient value of foods. *Critical Reviews in Food Science & Nutrition* 16(4):355–381.

Desikachar, H. S. R. 1975. Processing of maize, sorghum and millets for food uses. *Journal of Scientific and Industrial Research* 34:231–237.

Dhankher, N., and Chauhan, B. M. 1987. Effect of temperature and fermentation time on phytic acid and polyphenol content of Rabadi: a fermented pearl millet food. *Journal of Food Science* 52:828–829.

Dharmaraj, U., Ravi, R., and Malleshi, N. G. 2014. Cooking characteristics and sensory qualities of decorticated finger millet (eleusine coracana). *Journal of Culinary Science & Technology* 12:215–228.

Duhan, A., Chauhan, B. M., Punia, D., and Kapoor, A. C. 1989. Phytic acid content of chickpea (*Cicerarietinum*) and black gram (*Vignamungo*): varietal differences and effect of domestic processing and cooking methods. *Journal of the Science of Food and Agriculture* 49:449–455.

El-Hag, M. E., El-Tinay, A. H., and Yousif, N. E. 2002. Effect of fermentation and dehulling on starch, total polyphenols, phytic acid content and in vitro protein digestibility of pearl millet. *Food Chemistry* 77:193–196.

Elyas, S. H. A., El Tinay, A. H., Yousif, N. E., and Elshaikh, E. A. E. 2002. Effect of natural fermentation on nutritive and in-vitro protein digestibility of pearl millet. *Food Chemistry* 78:193–196.

Eyzaguirre, R. Z., Nienaltowska, K., De Jong, L. E., Hasenack, B. B., and Nout, M. J. 2006. Effect of food processing of pearl millet (*Pennisetum glaucum*) IKMP-5 on the level of phenolics, phytate, iron and zinc. *Journal of the Science of Food and Agriculture* 86(9):1391–1398.

Gabrelibanos, M., Tesfay, D., Raghavendra, Y., and Sintayeyu, B. 2013. Nutritional and health implication of legumes. *International Journal of Pharmaceutical Science and Research* 4:1269–1279.

Gahalawat, P., and Sehagal, S. 1992. Phytic acid, saponin and polyphenol in weaning foods prepared from oven heated green gram and cereals. *Cereal Chemistry* 69:463–464.

García-Estepa, R. M., Guerra-Hernández, E., and García-Villanova, B. 1999. Phytic acid content in milled cereal products and breads. *Food Research International* 32:217–221.

Gupta, A., Seetharam, A., and Mushonga, J. N. 1994. *Advances in small millets*, ed. Ken W. Riley. New York: International Science.

Gupta, C., and Sehgal, S. 1991. Development, acceptability and nutritional value of weaning mixtures. *Plant Foods for Human Nutrition* 41:107–116.

Hadimani, N. A., and Malleshi, N. G. 1993. Studies on milling, physico-chemical properties, nutrient composition and dietary fiber content of millets. *Journal of Food Science and Technology* 30:17–20.

Hassan, A. B., Ahmed, I. A., Osman, N. M., Eltayeb, M. M., Osman, G. A., and Babiker, E. E. 2006. Effect of processing treatments followed by fermentation on protein content and digestibility of pearl millet (*Pennisetum typhoideum*) cultivars. *Pakistan Journal of Nutrition* 5(1):86–89.

Hegde, P. S., and Chandra, T. S. 2005. ESR spectroscopic study reveals higher free radical quenching potential in kodo millet (*Paspalum scrobiculatum*) compared to other millets. *Food Chemistry* 92:177–182.

Heiniö, R. L., Katina, K., Wilhelmson, A., Myllymäki, O., Rajamäki, T., and Latva-Kala, K. 2003. Relationship between sensory perception and flavour-active volatile compounds of germinated, sourdough fermented and native rye following the extrusion process. *LWT-Food Science and Technology* 36:533–545.

Hithamani, G., and Srinivasan, K. 2014. Effect of domestic processing on the polyphenol content and bioaccessibility in finger millet (*Eleusine coracana*) and pearl millet (*Pennisetum glaucum*). *Food Chemistry* 164:55–62.

Hole, A. S., Rud, I., Grimmer, S., Sigle, S., Narvhus, J., and Sahlstrøm, S. 2012. Improved bioavailability of dietary phenolic acids in whole grain barley and oat groat following fermentation with probiotic. *Lactobacillus acidophilus, Lactobacillus johnsonii, and Lactobacillus reuteri*. *Journal of Agricultural and Food Chemistry* 60:6369–6375.

Jalgaonkar, K., Jha, S. K., and Sharma, D. K. 2016. Effect of thermal treatments on the storage life of pearl millet (*Pennisetum glaucum*) flour. *Indian Journal of Agricultural Sciences* 86(6):762–767.

Jaybhaye, R. V., Pardeshi, I. L., Vengaiah, P. C., and Srivastav, P. P. 2014. Processing and technology for millet based food products: a review. *Journal of Ready to Eat Food* 1(2):32–48.

Jha, N., Krishnan, R., and Meera, M. S. 2015. Effect of different soaking conditions on inhibitory factors and bioaccessibility of iron and zinc in pearl millet. *Journal of Cereal Science* 66:46–52.

Kachare, D. P., and Chavan, J. K. 1992. Effect of seed treatment on the changes in acidity of pearl millet meal during storage. *Indian Journal of Agricultural Biochemistry* 5:15–20.

Kapoor, R., and Kapoor, A. C. 1990. Effect of different treatments on keeping quality of pearl millet flour. *Food Chemistry* 60:189–191.

Kataria, A., Chauhan, B. M., and Punia, D. 1992. Digestibility of proteins and starch (in vitro) of amphidiploids (black gram × mung bean) as affected by domestic processing and cooking. *Plant Foods for Human Nutrition* 42:117–125.

Katina, K., Laitila, A., Juvonen, R., Liukkonen, K. H., Kariluoto, S., and Piironen, V. 2007. Bran fermentation as a means to enhance technological properties and bioactivity of rye. *Food Microbiology* 24:175–186.

Kaukovirta-Norja, A., Wilhelmson, A., and Poutanen, K. 2004. Germination: a means to improve the functionality of oat. *Agricultural and Food Science* 13:100–112.

Kebakile, M. M., Rooney, L. W., and Taylor, J. R. 2007. Effects of hand pounding, abrasive decortication-hammer milling, roller milling, and sorghum type on sorghum meal extraction and quality. *Cereal Foods World* 52:129–137.

Khetarpaul, N., Chauhan, B. M. 1989. Effect of germination and pure culture fermentation on HCl extractability of, minerals of pearl millet (*Pennisetum glaucum*). *International Journal of Food Science and Techology* 24:327–331.

Krishnan, R., Dharmaraj, U., and Malleshi, N. G. 2012. Influence of decortication, popping and malting on bioaccessibility of calcium, iron and zinc in finger millet. *LWT-Food Science and Technology* 48:169–174.

Lestienne, I., Besanon, P., Caporiccio, B., Lullien-Pellerin, V., and Treche, S. 2005. Iron and zinc in vitro availability in pearl millet flours (*Pennisetum glaucum*) with varying phytate, tannin, and fiber contents. *Journal of Agricultural and Food Chemistry* 53:3240–3247.

Lestienne, I., Buisson, M., Lullien-Pellerin, V., Picq, C., and Trèche, S. 2007. Losses of nutrients and anti-nutritional factors during abrasive decortication of two pearl millet cultivars (*Pennisetum glaucum*). *Food Chemistry* 100:1316–1323.

Lestienne, I., Icard-Vernière, C., Mouquet, C., Picq, C., and Trèche, S. 2005. Effects of soaking whole cereal and legume seeds on iron, zinc and phytate contents. *Food Chemistry* 89(3):421–425.

Lestienne, I., Mouquet-Rivier, C., Icard-Verniere, C., Rochette, I., and Trèche, S. 2005. The effects of soaking of whole, dehulled and ground millet and soybean seeds on phytate degradation and Phy/Fe and Phy/Zn molar ratios. *International Journal of Food Science and Technology* 40:391–399.

Liu, J., Tang, X., Zhang, Y., and Zhao, W. 2012. Determination of the volatile composition in brown millet, milled millet and millet bran by gas chromatography/mass spectrometry. *Molecules* 17:2271–2282.

Liukkonen, K. H., Katina, K., Wilhelmsson, A., Myllymaki, O., Lampi, A. M., Kariluoto, S., and Poutanen, K. 2003. Process induced changes on bioactive compounds in whole grain rye. *Proceedings of the Nutrition Society* 62:117–122.

Mahajan, S., and Chauhan, B. M. 1987. Phytic acid and extractable phosphorus of pearl millet flour as affected by natural lactic acid fermentation. *Journal of the Science of Food and Agriculture* 41:381–386.

Malleshi, N. G., Desikachar, H. S. R., and VenkatRao, S. 1986. Protein quality evaluation of a weaning food based on malted ragi and green gram. *Plant Foods for Human Nutrition* 36:223–230.

Malleshi, N. G., and Hadimani, N. A. 1993. Nutritional and technological characteristics of small millets and preparation of value-added products from them. In Riley, K. W. (ed). S.C.

Mallet, M., du Plessis, P., SA-DC, CRIAA. 2001. Mahagngu Post-Harvest Systems. Research Report prepared for Ministry of Agriculture. Water and Rural Development and Namibia Agronomic Board, Windhoek.

Mohamed, A. E., Mohamed Ahmed, I. A., and Babiker, E. E. 2010. Preservation of millet flour by refrigeration: changes in antinutrients, protein digestibility and sensory quality during processing and storage. *Research Journal of Agriculture and Biological Sciences* 6(4):411–416.

Mohamed, M. E., Amro, B. H., Mashier, A. S., and Elfadil, E. B. 2007. Effect of processing followed by fermentation on antinutritional factors content of pearl millet (*Pennisetum glaucum* L.) cultivars. *Pakistan Journal of Nutrition* 6 (5):463–467.

Murthy, K. V., Ravi, R., Bhat, K. K., and Raghavarao, K. S. M. S. 2008. Studies on roasting of wheat using fluidized bed roaster. *Journal of Food Engineering* 89:336–342.

Nagah, A., and Seal, C. 2005. In vitro procedure to predict apparent antioxidant release from wholegrain foods measured using three different analytical methods. *Journal of the Science of Food and Agriculture* 85:1177–1185.

Naikare, S. M., Chavan, J. K., and Kadam, S. S. 1986. Depigmentation and utilization of pearl millet in the preparation of cookies and biscuits. *J Maharashtra AgricUniv* 11:90–93.

Nantanga, K. K. M., Seetharaman, K., Kock, H. L., and Taylor, J. R. N. 2008. Thermal treatments to partially pre-cook and improve the shelf life of whole pearl millet flour. *Journal of the Science of Food and Agriculture* 88:1892–1899.

Nibedita, M., and Sukumar, B. 2003. Extrusion cooking technology employed to reduce the anti-nutritional factor tannin in sesame meal. *Journal of Food Engineering* 56:201–202.

Obadina, A., Ishola, I. O., Adekoya, I. O., Soares, A. G., de Carvalho, C. W. P., and Barboza, H. T. 2016. Nutritional and physico-chemical properties of flour from native and roasted whole grain pearl millet (*Pennisetum glaucum* [L.] R. Br.). *Journal of Cereal Science* 70:247–252.

Obizoba, I. C., and Atii, J. V. 1994. Evaluation of the effect of processing techniques on the nutrient and antinutrient contents of pearl millet (*Pennisetum glaucum*) seeds. *Plant Food for Human Nutrition* 45:23–34.

Ocheme, O. B., and Chinma, C. E. 2008. Effects of soaking and germination on some physicochemical properties of millet flour for porridge production. *Journal of Food Technology* 6(5):185–188.

Opoku, A. R., Ohenhen, S. O., and Ejiofor, N. 1981. Nutrient composition of millet (*Pennisetum typhoides*) grains and malts. *Journal of Agricultural and Food Chemistry* 29:1247–1248.

Parada, J., and Aguilera, J. 2007. Food microstructure affects the bioavailability of several nutrients. *Journal of Food Science* 72:R21–R32.

Pawar, V. D., and Machewad, G. M. 2006. Processing of foxtail millet for improved nutrient availability. 30:269–279.

Pawar, V. D., and Parlikar, G. S. 1990. Reducing the polyphenols and phytate and improving the protein quality of pearl millet by dehulling and soaking. *Journal of Food Science and Technology* 27:140–143.

Pellegrini, N., Chiavaro, E., Gardana, C., Mazzeo, T., Contino, D., and Gallo, M. 2010. Effect of different cooking methods on color, phytochemical concentration, and antioxidant capacity of raw and frozen Brassica vegetables. *Journal of Agricultural and Food Chemistry* 58:4310–4321.

Prom-u-thai, C., Huang, L., Glahn, R., Welch, R., Fukai, S., and Rerkasem, B. 2006. Iron (Fe) bioavailability and the distribution of anti-Fe nutrition biochemicals in the unpolished, polished grain and bran fraction of five rice genotypes. *Journal of the Science of Food and Agriculture* 86:1209–1215.

Rai, K. N., Gowda, C. L. L., Reddy, B. V. S., and Sehgal, S. 2008. Adaptation and potential uses of sorghum and pearl millet in alternative and health foods. *Comprehensive Reviews in Food Science and Food Safety* 7:340–352.

Reyden, P., and Selvendran, R. R. 1993. Phytic acid: properties and determination. In Macrae, R., Robinson, R. K., and Sadler, M. J. (eds). *Encyclopedia of food science, food technology and nutrition* pp. 3582–3587. London: Academic Press.

Rohrbach, D. D., and Obilana, A. B. 2004. The commercialization of sorghum and pearl millet in Africa: traditional and alternative foods, products and industrial uses in perspective. In *Alternative uses of Sorghum and Pearl Millet in Asia: proceedings of the expert meeting*, 2003 July 1–4. Patancheru, Andhra Pradesh: ICRISAT. CFC Technical Paper Nr 34, pp. 233–263.

Saharan, K., Khetarpaul, N., and Bishnoi, S. 2001. HCl-extractability of minerals from rice bean and faba bean: influence of domestic processing methods. *Innovative Food Science and Emerging Technologies* 2(4):323–325.

Salar, R. K., Purewal, S. S., and Bhatti, M. S. 2016. Optimization of extraction conditions and enhancement of phenolic content and antioxidant activity of pearl millet fermented with Aspergillus awamori MTCC-548. *Resource-Efficient Technologies* 2(3):148–157.

Salar, R. K., Purewal, S. S., and Sandhu, K. S. 2017. Fermented pearl millet (*Pennisetum glaucum*) with in vitro DNA damage protection activity, bioactive compounds and antioxidant potential. *Food Research International* 100:204–210.

Saleh, A. S., Zhang, Q., Chen, J., and Shen, Q. 2013. Millet grains: nutritional quality, processing, and potential health benefits. *Comprehensive Reviews in Food Science and Food Safety* 12(3):281–295.

Sandhu, K. S., Godara, P., Kaur, M., and Punia, S. 2017. Effect of toasting on physical, functional and antioxidant properties of flour from oat (*Avena sativa* L.) cultivars. *Journal of the Saudi Society of Agricultural Sciences* 16(2):197–203.

Sandhu, K. S., and Punia, S. 2017. Enhancement of bioactive compounds in barley cultivars by solid substrate fermentation. *Journal of Food Measurement and Characterization* 11(3):1355–1361.

Sandhu, K. S., Punia, S., and Kaur, M. 2016. Effect of duration of solid state fermentation by Aspergillus awamori nakazawa on antioxidant properties of wheat cultivars. *LWT-Food Science and Technology* 71:323–328.

Sandhu, K. S., Punia, S., and Kaur, M. 2017. Fermentation of cereals: a tool to enhance bioactive compounds. In *Plant biotechnology: recent advancements and developments* pp. 157–170. Singapore: Springer.

Scheuring, J. F., and Rooney, L. W. 1979. A staining procedure to determine the extent of bran removal from pearled sorghum. *Cereal Chemistry* 56:545–548.

Serna-Saldivar, S. O., Clegg, C., and Rooney, L. W. 1994. Effects of parboiling and decortication on the nutritional value of sorghum (*Sorghum bicolor L. Moench*) and pearl millet (*Pennisetum glaucum L.*). *Journal of Cereal Science* 19:83–89.

Sharma, P., Gujral, H. S., and Singh, B. 2012. Antioxidant activity of barley as affected by extrusion cooking. *Food Chemistry* 131(4):1406–1413.

Siddhuraju, P., and Becker, K. 2001. Effect of various domestic processing methods on antinutrients and in vitro-protein and starch digestibility of two indigenous varieties of Indian pulses. *Mucuna pruries* var*utilis*. *Journal of Agriculture and Food Chemistry* 49(3):058–67.

Sihag, M. K., Sharma, V., Goyal, A., Arora, S., and Singh, A. K. 2015. Effect of domestic processing treatments on iron, β-carotene, phytic acid and polyphenols of pearl millet. *Cogent Food & Agriculture* 1(1):1109171.

Silva, C. A. B., Rayes, R. P., and Deng, J. C. 1981. Influence of soaking and cooking upon softening and eating quality of black bean (*Phaseolus vulgaris*). *Journal of Food Science* 46:1716–1720.

Singh, A. K., Rehal, J., Kaur, A., and Jyot, G. 2015. Enhancement of attributes of cereals by germination and fermentation: a review. *Critical Reviews in Food Science and Nutrition* 55:1575–1589.

Singh, P., and Raghuvanshi, R. S. 2012. Finger millet for food and nutritional security. *African Journal of Food Science* 6:77–84.

Siroha, A. K., and Sandhu, K. S. 2017. Effect of heat processing on the antioxidant properties of pearl millet (*Pennisetum glaucum* L.) cultivars. *Journal of Food Measurement and Characterization* 11(2):872–878.

Slavin, J. L., Martini, M. C., Jacobs, D. R., and Marquart, L. 1999. Plausible mechanisms for the protectiveness of whole grains. *The American Journal of Clinical Nutrition* 70:459S–463S.

Suma, P. F., and Urooj, A. 2014. Influence of germination on bioaccessible iron and calcium in pearl millet (*Pennisetum typhoideum*). *Journal of Food Science and Technology* 51(5):976–981.

Sumathi, A., Ushakumari, S. R., and Malleshi, N. G. 2007. Physico-chemical characteristics, nutritional quality and shelf-life of pearl millet based extrusion cooked supplementary foods. *International Journal of Food Sciences and Nutrition* 58 (5):350–362.

Sushma, D., Yadav, B. K., and Tarafdar, J. C. 2008. Phytate phosphorus and mineral changes during soaking, boiling and germination of legumes and pearl millet. *Journal of Food Science and Technology* 45(4):344–348.

Taylor, J. R. N. 2004. Millet: Pearl. In Wrigley, C., Corke, H., and Walker, C. E. (eds). *Encyclopedia of grain science* pp. 253–261. Vol 2. London: Elsevier.

Tian, B., Xie, B., Shi, J., Wu, J., Caia, Y., Xu, T., Xue, S., and Deng, Q. 2010. Physico-chemical changes of oat seeds during germination. *Food Chemistry* 119:1195–1200.

Tiwari, A., Jha, S. K., Pal, R. K., and Sethi, K. L. 2014. Effect of pre-milling treatments on storage stability of pearl millet flour. *Journal of Food Processing and Preservation* 38:1215–1223.

Towo, E. E., Svanberg, U., and Ndossi, G. D. 2003. Effect of grain pre-treatment on different extractable phenolic groups in cereals and legumes commonly consumed in Tanzania. *Journal of the Science of Food and Agriculture* 83:980–986.

Wang, T., He, F., and Chen, G. 2014. Improving bioaccessibility and bioavailability of phenolic compounds in cereal grains through processing technologies: a concise review. *Journal of Functional Foods* 7:101–111.

Yadav, D. N., Anand, T., Kaur, J., and Singh, A. K. 2012. Improved storage stability of pearl millet flour through microwave treatment. *Agricultural Research* 1 (4):399–404.

Zielinski, H., Michalska, A., Piskula, M. K., and Kozlowska, H. 2006. Antioxidants in thermally treated buckwheat groats. *Molecular Nutrition and Food Research* 50:824–832.

# 5 Starch

## Structure, Properties and Applications

*Anil Kumar Siroha and Sneh Punia*

## CONTENTS

## 5.1 INTRODUCTION

Starch is the most abundant reserve carbohydrate found in plants and is a major source of energy in human food. It is a main component of various foods and its properties and reaction with other food components, mainly water and lipids, are important to the food industry and for human nutrition (Copeland et al., 2009). Therefore, a detailed investigation is necessary to analyse the biochemical and functional properties of starches (Kaur et al., 2004). Pearl millet has protein, fat, crude fiber and carbohydrate content: 9.7–11.3%, 5.1–7.2%, 2.9–3.8% and 69.6–72.5%, respectively (Siroha et al., 2016). Starch is the principal carbohydrate constituent of a pearl millet grain, and has been reported at 62.8–70.5% for different Indian cultivars by Suma and Urooj (2015). Starch is made of mainly two components, amylose and amylopectin; amylose is made from α-(1/4) D-glucopyranosyl units while amylopectin is made up of small chains which are higher when compared to amylose linked together at their reducing end side by a α-(1/6) linkage (Biliaderis,

1998). Rheological characteristics of starches depend upon amylose, amylopectins and functional groups present in the starches, that is, the phosphate group (Berski et al., 2011).

Pearl millet is an underutilized crop and it is mainly used for animal feed and fodder. It has about 60–70% starch content, which is isolated easily and may be used for various food industries. Food industries are dependent on three main staple food crops: corn, rice and potato for starch, and by isolating the starch from pearl millet the reliance on those crops for it may be reduced. Understanding the properties, structure, and applications of pearl millet starch significantly contributes to the further utilization of pearl millets as an alternative functional crop. This chapter summarizes the structural, morphological, *in vitro* digestibility and rheological properties of pearl millet starches and this provides direction for new research.

## 5.2  STARCH ISOLATION

Wet milling methods are used to separate the starch from pearl millet grains or flour which have been steeped in aqueous solutions for few hours. After the grains have softened, they are milled to slurry. The slurry is passed through different sieves to separate protein and other fibrous matter. After sieving, centrifuging is done to remove the remaining protein and fibrous matter. Crude starch is repeatedly washed with water and centrifuged to obtain pure starch. The starch is then dried in a conventional oven at a temperature of 40–50°C (Khatkar et al., 2013; Sandhu and Singh, 2005; Wu et al., 2014). Balasubramanian et al. (2014) used NaOH and $NaHSO_3$ for steeping the flour to isolate the starch. Wu et al. (2014) followed an alkaline steeping method to isolate the starch, which was frequently washed with 0.2% (w/v) sodium hydroxide and 0.5% (w/v) sodium dodecyl sulfonate to discard any remaining protein. Suma and Urooj (2015) isolated starch by using the method described by Sowbhagya and Bhattacharya (1971). Sodium azide is used to inhibit the bacterial growth and crude starch is purified by suspending in NaCl (0.1 N): Toluene (1:1) followed by centrifugation and washing with distilled water, alcohol and acetone. Zhong et al. (2009) observed that method used to isolate starch affects the pasting and rheological properties. The starch isolation from pearl millet grains is shown in Flow diagram 5.1.

500g whole grains (10–20% moisture content)
↓
Steeping (distilled water (1.25 litre+ $Na_2S_2O_5$ (0.1%)
↓ at 50°C for 18–20 h
Grinding of grains
↓
Sieving of slurry through 45, 75, 100, 150, and 250 mesh
↓
Starch protein slurry allowed to stand for 4–5 h
↓
Removal of supernatant and re-suspension of settled starch layer in water
↓

Centrifugation at 3000 rpm for ten min

↓

Scrapping of upper layer and re-suspension of white layer in water 3–4 times

↓

Drying in an oven at 45°C for 12 h

**Flow diagram 5.1 Separation of starch from grains** (Sandhu and Singh, 2005)

## 5.3 CHEMICAL COMPOSITION

Starch is semi-crystalline in nature with varying levels of crystallinity. The crystallinity is exclusively associated with the amylopectin component, while the amorphous regions mainly represent amylose (Zobel, 1988a, 1988b). Significant variations were found for the chemical composition of pearl millet starches (Table 5.1). The amylose content of pearl millet starches is observed to be 28.8–31.9% (Hoover et al., 1996), 15.64–19.46% (Khatkar et al., 2013), 28.14% (Wu et al., 2014), 32.5% (Annor et al., 2014a), 3.96–4.96% (Suma and Urooj, 2015) and 13.6–18.1% (Sandhu and Siroha, 2017). Singh et al. (2006) reported that the amylose content of starches is affected by the botanical source, climate and conditions of soils during grain maturity. Zhu (2016) stated that if different methods are used to analyse the same sample, this may give different results. For example, when amylose content is measured with an iodine solution, some part of iodine will react with the amylopectin unit and give higher values for amylose content (Yoshimoto et al., 2004). Hoover (2001) stated that if starch is not defatted prior to amylose estimation, the amylose content will be underestimated.

Starch mainly contains amylose and amylopectin content, with other minor components such as protein, fat, fiber and ash also being present (Table 5.1). The purity of starches is based on the quantity of minor components, lower

**TABLE 5.1**

**Starch yield and chemical composition of native pearl millet starches**

| Starch yield % | Amylose content % | Protein % | Fat % | Ash % | References |
|---|---|---|---|---|---|
| 42.7–47.5 | 13.6–18.1 | 0.32–0.75 | 0.27–0.46 | 0.22–0.41 | Sandhu and Siroha, 2017 |
| 34.50–39.40 | 3.96–4.96 | 0.53–0.55 | 0.37–0.38 | 0.10–0.10 | Suma and Urooj, 2015 |
| | 28.14 | 0.31 | | 0.86 | Wu et al., 2014 |
| 70.4 | 32.5 | | | | Annor et al., 2014a |
| | 15.64–19.46 | | 0.119–0.23 | 0.11–0.12 | Khatkar et al., 2013 |
| 53.1 | | 2.04 | | 1.35 | Balasubramanian et al., 2014 |
| | 6.0 | 0.2 | 0.2 | 1.4 | Choi et al., 2004 |
| 53.1–56.5 | 28.8–31.9 | | | 0.02–0.03 | Hoover et al., 1996 |
| 60.2 | 22.8 | 0.68 | 0.92 | 0.78 | Wankhede et al. (1990) |

values showing higher purity of starches. These components differ with regard to starch isolation as well as quantification methods. Estrada-León et al. (2016) reported that the protein content of the starches is affected by the method of extraction, with the sodium bisulphite and acid steeping methods removing more protein from the starches in comparison with the distilled water method due to the solvents used in those methods, which facilitated the solubility of the proteins and their removal by centrifugation. Zhong et al. (2009) also reported that an isolation method affects the protein content of starch; they observed that 0.4% NaOH and SDS treatments appeared to produce starches with slightly lower residual protein contents.

## 5.4   CHEMICAL STRUCTURE

Starch molecules are categorized into amylose and amylopectin. Amylose is linear with the glucosyl units, being connected by α-(1−4)-linkages. A few branches of α-(1−6)-linkages may exist in amylose. Amylopectin has a highly branched structure, in which the glucosyl units are branched by α-(1−6)-bonds (Bertoft, 2004). Amylose has few branches, whereas amylopectin is larger with many branches. The branches in amylopectin are arranged in a clustered fashion. The tightly branched unit in the amylopectin is termed a building block, which has an internal chain length of 1–3 glucosyl residues (Pérez & Bertoft, 2010). These two biopolymers form crystalline and amorphous areas in the starch granule on multiple scales to form a semi-crystalline structural system (Lin et al., 2016; Qiao et al., 2017). On the basis of size distribution of the internal unit chains, the structure of amylopectin was divided into four types. Type 1 amylopectins have typically very few long B-chains and a broad size distribution of short B-chains. Type 2 has considerably more long B-chains and a narrower size distribution of the short chains. Type 3 has a few more long chains compared to Type 2, whereas Type 4 has considerably more long B-chains (Bertoft et al., 2008). Starch polymer behavior depends on measurable parameters, such as molecular weight distribution, the nature of chain branching, and chain lengths (Weaver et al., 1988). The molecular weight of amylose is approximately $1 \times 10^6$ g mol$^{-1}$, and the molecule exhibits a degree of polymerization (DP) of 250–1000 D-glucose units, whereas amylopectin is one of the largest molecules found in nature, with a molecular weight in the order of $1 \times 10^7$ to $1 \times 10^9$ g mol$^{-1}$ and a DP of 5000–50,000 D-glucose units (Pérez et al., 2009). Gaffa et al. (2004) reported the degree of polymerization (DP: the number of glucose monomer units in the chain) by number for amylose and amylopectin, 1060–1250 and 9000–9100, and average chain length for amylose and amylopectin was observed at 260–270 and 20–21, respectively, for pearl millet starches. The chains can then be grouped into short (DP <36) and long chains (DP >36) (Bertoft, 2004). Pérez and Bertoft (2010) stated that chains of amylopectin can be classified as internal and external chain: external chains are found in the crystalline parts of the amylopectin molecule, whereas the internal chains are generally considered to be in the amorphous part. Annor et al. (2014a) observed 32.5% amylose content for pearl millet starch, in which long chains were observed at 15.8% and short chain at 16.7%, where the ratio of long chains and short chains was found to be 0.95. Annor et al. (2014b) found average chain length, short chain length, long chain length, external chain length and internal chain length to be 18.0, 15.80, 52.12, 11.95, 5.0, respectively, for pearl millet starch. Gaffa et al. (2004) also observed values of 13.2–13.8 and 5.8–6.2 for external and internal chain for amylopectin of pearl millet starches. Pasting properties of starch are affected by amylose and lipid contents and by branch chain length distribution of amylopectin (Jane et al., 1999). Jane and Chen (1992) reported that the viscosity of starch paste is affected by the chain length of amylopectin and the molecular size of amylose.

## 5.5 PHYSICOCHEMICAL PROPERTIES

Swelling power and water solubility provide measures of the magnitude of inter-action between starch chains within the amorphous and crystalline domains (Lin et al., 2016). The extent of this interaction is influenced by the contents and charac-teristics of amylose and amylopectin (Kaur et al., 2007). The swelling power of starch is related to the gelatinization of starch, reflecting the breaking of hydrogen bonds in the crystalline regions, uptake of water by hydrogen bonds and water absorption by non-starch polysaccharides and protein (Thitipraphunkul et al., 2003; Yu et al., 2012). Swelling power is positively correlated with amylopectin short branch chains and negatively with amylopectin long branch chains (Salman et al., 2009). The significantly negative correlation of amylose with swelling power and significantly positive correlation with water solubility have been reported in some starches (Chung et al., 2011; Singh et al., 2006). Amylose restrains swelling and maintains the integrity of swollen granules, and the lipid-complexed amylose chains restrict both granular swelling and amylose leaching (Tester and Morrison, 1992). Swelling power and solubility of the starch granule is influenced by many factors, including amylose to amylopectin ratio and contents, the molecular mass of each fraction, degree of branching, conformation length of the outer branch of amylopectin and the presence of other components such as lipids and proteins (Mauro, 1996). Falade and Okafor (2013) stated that large granules swell more rap-idly than small granules when starch was heated in water so that larger starch gran-ules display higher swelling power. Hoover et al., (1996) reported swelling power and solubility of pearl millet starches in the range of 18.0–28.6 g/g and 9.45–10.40 g/100 g, respectively. Swelling power and solubility of starches from different culti-vars ranged from 14.1 to 17.9 g/g and 10.4–16.2 g/100 g, respectively (Sandhu and Siroha, 2017). The starch granules are discrete semi-crystalline aggregates consist-ing of amylose and amylopectin as major components. The ratio of these fractions in the starch granule and the manner in which they are arranged inside the granule affect the swelling and solubility of the starch (Beleia et al., 1980).

## 5.6 MORPHOLOGICAL PROPERTIES

Scanning electron microscopy (SEM) helps to analyze the interrelation between starch granule morphology and starch genotype (Fannon et al., 1992). The morphological characteristics of starches from different plant sources vary according to the genotype and biological origin. The size and shape of starch granules varied with biological origin (Svegmark and Hermansson, 1993). Starch granules range in size (1–100 µm diameter) and shape (polygonal, spherical, lenticular), and can differ in content, structure and organization of the amylose and amylopectin molecules, the branching pattern of amylopectin, and the level of crystallinity (Lindeboom et al., 2004). Usually, the variations in external morphology are sufficient to establish the characterization of botanical origin through optical microscopy (Pérez and Bertoft, 2010). Starch granule size is reported to affect its physico-chemical properties, such as crystallinity, solubility, pasting and enzyme susceptibility (Lindeboom et al., 2004). The complexity of starch biosynthesis results in natural variability in amylose and amylopectin molecules, which is reflected in the diversity of granules (Wani et al., 2013). The starch from different botanical origins appeared to have particular shapes and characteristic dimensions (Singh et al., 2003). Conversely, the morphological properties of starch could be an index for estimating the botanical origin (Wang et al., 2012). The variation in granule size distribution and shape is reported to be associated with functional properties in different food systems (Peterson and Fulcher, 2001). The granule shape and size of pearl millet starch are reported in Table 5.2 and shown in Figure 5.1. Pearl millet starch granules show small to large, round, spherical and polygonal granules with some pores and indentations (Annor et al., 2014a; Sandhu and Siroha, 2017; Shaikh et al., 2015; Wu et al., 2014). Various researchers reported granule size of pearl millet starch of 3–23 µm (Annor et al., 2014a; Shaikh et al., 2015; Wu et al., 2014).

## TABLE 5.2
## Morphology of starch granules

| Shape of starch granules | Size (µm) | References |
|---|---|---|
| Small to large, spherical and polygonal | | Sandhu and Siroha, 2017 |
| Polygonal shape with a few spherical shaped granules with indentations and creases | 6–12 | Shaikh et al., 2015 |
| Large polygonal and few small spherical granules | 3–15 | Wu et al., 2014 |
| Polygonal in shape with a few spherical granules | 3.5–23 | Annor et al., 2014a |
| Polygonal in shape with round edges and some pores at the surface | 5–10 | Choi et al., 2004 |
| Polygonal or round in shape having deep indentations and pores | 10.5 | Hoover et al., 1996 |
| Polygonal or round in shape with deep indentations | 10–15.5 | Wankhede et al., 1990 |
| Spherical to polygonal having deep indentations | 8–12 | Badi et al., 1976 |

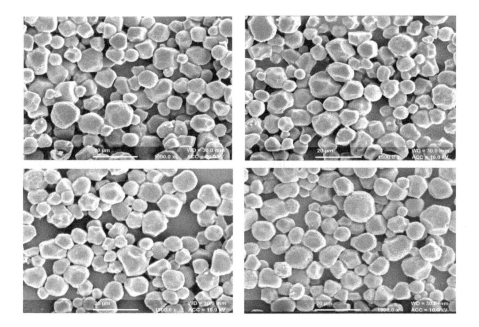

**FIGURE 5.1** Scanning electron micrographs of starches from different pearl millet cultivars

## 5.7 X-RAY DIFFRACTION PATTERN

X-ray diffraction (XRD) is used to study the presence and characteristics of the crystalline structure of the starch granules. Starch granules possess a semi-crystalline structure corresponding to different polymorphic forms and based on this, starch can be classified into three types, namely A, B and C (Buleon et al., 1998). Arrangement of the amylopectin helices are supposed to be responsible for starch crystallinity, whereas amylose is associated to amorphous regions (Singh et al., 2006; Zobel, 1988a). The crystalline parts of starch always show sharp peaks, whereas amorphous parts of starch are diffuse (Gernat et al., 1990). On the basis of these peaks, starch is differentiated into A, B and C types. Differences in diffraction patterns may be due to different growth conditions and maturity of the parent plant at the time of harvest, biological origin, amylose and amylopectin content (Zhou et al., 2010).

Diffraction pattern for pearl starch showed A-type diffraction patterns with peaks at 15°, 17°, 18° and 23.1° (2θ), which is a characteristic of cereal starches (Sandhu and Siroha, 2017). Suma and Urooj (2015) also reported A-type diffraction patterns for Indian pearl millet varieties (Kalukombu and Maharashtra Rabi Bajra) with sharp peaks at 15° and 23° and diffused peaks at 17° and 18° (2θ). Wu et al., (2014) reported the crystalline degree of 30.89% for pearl millet starch. The differences in relative crystallinity among starches can be attributed to the following: (1) crystal size, (2) amount of crystalline regions (influenced by amylopectin content and amylopectin chain length), (3) orientation of the double helices within the crystalline domains, and (4) extent of interaction between double helices (Hoover and Ratnayake, 2002).

## 5.8 RHEOLOGICAL PROPERTIES

### 5.8.1 PASTING PROPERTIES

Pasting properties of starch can be measured using an amylograph, such as the Brabender amylograph and rapid visco-analyzer (RVA) (Hoover et al., 1996; Sandhu and Singh, 2005; Sandhu et al., 2007; Shaikh et al., 2015), or using a dynamic rheometer (Punia et al., 2019a, 2019b; Siroha and Sandhu, 2018; Siroha, Sandhu, Kaur et al., 2019; Siroha, Sandhu and Punia, 2019). Pasting properties are shown in Table 5.4 and Figure 5.2. Pasting is a complex phenomenon which specifically refers to the changes in starch after post-gelatinization heating (Wani et al., 2016). When starch is heated in excess water while being stirred, irreversible swelling of starch granules occurs; this is accompanied by leaching of linear amylose molecules and possibly solubilization of branched chain amylopectin molecules, resulting in the formation of starch paste (Gani et al., 2017). It includes further swelling and polysaccharide leaching from the starch granules that increase viscosity due to the application of shear force (Atwell et al., 1988; Tester and Morrison, 1990). PV is regarded as the maximum viscosity attained by the sample and the tendency of starch granules to swell freely before physical

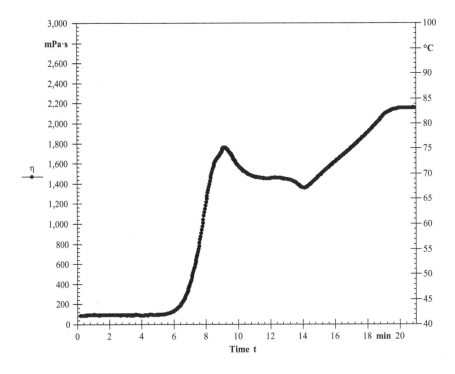

**FIGURE 5.2**   Pasting graph of pearl millet starch

breakdown (Jan et al., 2016). PV was found to be 1665–1998 mPa.s, 2826 mPa.s and 4647–8303 mPa.s for pearl millet starches (Khatkar et al., 2013; Sandhu and Siroha, 2017; Sharma et al., 2016). The increase in viscosity with an increase in temperature may be accredited to the removal of water from the exuded amylose of granules as they swell (Ghiasi et al., 1982). Breakdown viscosity (BV) is indicator of the minimum value of the apparent viscosity of the sample at the end of the isothermal region of the pasting profile. BV of pearl millet starch was observed to be 414–769 cP, 1313 mPa.s and 2273–3344 mPa.s, respectively (Khatkar et al., 2013; Sandhu and Siroha, 2017; Sharma et al., 2016). The breakdown is caused by the disintegration of the gelatinized starch granule structure during continuous stirring and heating (Whistler and BeMiller, 1997) According to Lee et al. (1995), breakdown viscosity is influenced by amylose content. The setback viscosity of starch is the difference between the final viscosity and trough viscosity. Setback viscosity of pearl millet starches was observed to be 8502 cP, 2439 mPa.s and 1342–4722 mPa.s, respectively (Khatkar et al., 2013; Sandhu and Siroha, 2017; Sharma et al., 2016). Setback reflects the degree of retrogradation. Pasting temperature is the minimum temperature required to cook the starch.

Hoover et al. (1996) observed pasting temperatures of 89.3–90°C for pearl millet starches, while Sandhu and Siroha (2017) reported pasting temperatures of 72.4–73.9°C for starches from different pearl millet cultivars. The pasting temperature can probably be affected by factors such as the degree of branching of amylopectin and a higher degree of crystallinity (Kim et al., 1996; Wang et al., 2011).

## 5.8.2 FLOW PROPERTIES

Many researchers have used rheological methods to determine the visco-elasticity of the starch pastes (Jan et al., 2016; Kaur and Singh, 2016; Kaur et al., 2004; Sandhu et al., 2004; Siroha and Sandhu, 2018). The most common method of studying the visco-elastic properties of starch is by means of a dynamic rheometer. G′ is the amount of energy stored in material and recovered from it per cycle, while G″ is the amount of energy dissipated or lost per cycle of sinusoidal deformation (Stanley et al., 1996). G′ and G″ values for starch suspensions, increased to a maximum and then decreased with continuous heating. The increase in G′ during heating may be due to starch granules swelling, followed by leaching. After attaining peak viscosity, G′ decreased for starches, indicating that their gel structure was destroyed during prolonged heating (Tsai et al., 1997). Starch granules swell when heated in excess water and their volume fraction and morphology play important roles in the rheological behavior of starch dispersions (Bagley & Christianson, 1982; Da Silva et al., 1997). The values of G′ and G″ ranged from 3613 to 6334 Pa and 352–441 Pa during heating (Sandhu and Siroha, 2017). Wu et al., (2014) reported that G′ and G″ values for pearl millet starch was not detectable for 6% starch concentration. The rheological properties are influenced by the starch concentration, amylose content, temperature, heating rate, and mechanical treatments. Furthermore, the addition and the presence of other components (lipids, proteins, sugars, salts, etc.) also influence these properties (Eliasson and Gudmundsson, 1996).

The frequency dependencies of the storage modulus (G') and loss modulus (G'') provide significantly valuable information about the gel structure (Kaur et al., 2008). In addition to G' and G'', the loss tangent (G''/G') was observed to reflect the dynamic elastic nature of gels, indicating the relative measure of the associated energy loss versus the energy stored per deformation cycle (Toker et al., 2013). The magnitude of G' and G'' of pearl millet starch pastes increased with an increase in angular frequency, with values of G' (997–1871 Pa) was much higher than G'' (67–107 Pa); the tan $\delta$ value of starch pastes was found to be less than 1, showing that pastes are more elastic than they are viscous (Sandhu and Siroha, 2017). Sharma et al. (2016) also reported G' and G'' values of 33.5 Pa and 4.5 Pa for pearl millet starch gels. Many applications of polymers are dependent upon their visco-elastic properties. For drug delivery purposes, the visco-elastic property of a polymer defines its application, especially in liquid drug delivery platforms (Nep et al., 2016). The large difference in the G' and G'' values of starch pastes between different studies is more likely due to the quantification method and concentration of the samples used for analysis, rather than the samples themselves. Therefore, the data from different studies are impossible to compare.

During steady shear properties, yield stress ($\sigma$o), consistency index (K) and flow behavior index (n) is determined. The yield stress value is indicative of the minimum force required to initiate the flow of starch paste. The yield stress value for pearl millet starch was 25.6–183.4 Pa (Sandhu and Siroha, 2017). Sharma et al. (2016) observed a 2.18 Pa yield stress for pearl millet starch. An n value less than 1 indicates shear thinning behavior of starch pastes. According to Morris (1989), the observed shear thinning behavior can be explained by the disruption of a network of entangled polysaccharide molecules during shearing. With increasing shear rate, the rate of disruption of the existing intermolecular entanglements becomes greater than the rate of reformation, consequently leading to the resultant reduction in apparent viscosity. The values of the flow behavior index, consistency index and yield stress are dependent on starch type, starch concentration and temperature (Wani et al., 2013). The large difference in $\sigma$o and K values of starch pastes between different studies is more likely due to the evaluating method and concentration of the sample used for analysis. Therefore, the data from different studies are also impossible to compare.

## 5.9 THERMAL PROPERTIES

Starch, when heated in the presence of excess water, undergoes an order–disorder phase transition called gelatinization over a temperature range characteristic of the starch source (Hoover et al., 2010). The thermal properties of pearl millet starches are shown in Table 5.3. The starch gelatinization temperature is a measure of the cooking quality of starch and an important parameter in food processing; starches with low gelatinization temperatures have good cooking quality (Waters et al., 2006). The differences in gelatinization temperatures may be attributed to the differences in amylose content, size, form and distribution of

## TABLE 5.3

### Thermal properties of pearl millet starches

| Starch:water ratio (w/w) | Heating rate (°C/min) | Gelatinization parameters | | | | References |
|---|---|---|---|---|---|---|
| | | $T_o$ (°C) | $T_p$ (°C) | $T_c$ (°C) | $\Delta Hgel$ (J/g) | |
| 1:2 | 5 | 63.4–67.7 | 69.3–71.6 | 74.5–76.3 | 10.6–12.4 | Sandhu and Siroha, 2017 |
| 1:4 | 10 | 59.9 | 62.6 | 73.6 | 11.9 | Sharma et al., 2016 |
| 1:3 | 10 | 69.6 | 73.7 | 87.2 | 9.8 | Shaikh et al., 2015 |
| 1: 2 | 10 | 64.7 | 69.7 | 74.6 | 3.5 | Wu et al., 2014 |
| 1:3 | 5 | 62.8 | 67.9 | | 12.3 | Annor et al., 2014a |
| 1:2 | 10 | 60.8–64.8 | 67.1–69.9 | 74.1–78.0 | 5.6–7.7 | Khatkar et al., 2013 |
| 1:3 | 10 | 65.3 | 76.4 | 81.6 | 8.5 | Choi et al., 2004 |
| 1:2 | 5 | 66.2–67.2 | 69.7–71.4 | 86.3–91.0 | 14.3–14.7 | Gaffa et al., 2004 |
| 1:3 | 10 | 60.9–64.5 | 67.5–70.0 | 74.0–78.0 | 2.5–3.5 | Hoover et al., 1996 |

starch granules, and to the internal arrangement of starch fractions within the granule (Singh et al., 2004). Bao et al. (2009) reported that the starch thermal properties are correlated with the chain length distribution and the average chain length of amylopectin molecules. The lower $\Delta H$ gel suggests a lower percentage of organized arrangements or a lower stability of the crystals (Chiotelli and Meste, 2002). According to Jenkins (1994) gelatinization in excess water is primarily a swelling driven process. This swelling acts to destabilize the amylopectin crystallites within the crystalline lamellae, which are ripped apart (smaller crystallites are destroyed first) during the process (Ratnayake et al., 2002). Krueger et al. (1987) reported that the higher the amylopectin content of the starch, the narrower was the temperature range of gelatinization. The $\Delta Hgel$ reflected primarily the loss of molecular double-helical order (Cooke and Gidley, 1992). $T_p$ gives a measure of crystallite quality (double helix length). Enthalpy of gelatinization ($\Delta Hgel$) gives an overall measure of crystallinity (quality and quantity) and is an indicator of the loss of molecular order within the granule (Cooke and Gidley, 1992; Hoover and Vasanthan, 1994; Tester and Morrison, 1990). The gelatinization transition temperatures $T_o$ (onset), $T_p$ (peak), $T_c$ (endset), and the enthalpy of gelatinization ($\Delta Hgel$) have been known to be influenced by the molecular architecture of the crystalline region (Wani et al., 2016). The starch to water ratio and the scanning rate of the temperature are influential to the gelatinization parameters recorded by DSC (Zhu, 2014). Sandhu and Siroha (2017) reported transition temperatures ($T_o$, $T_p$ and $T_c$) of pearl millet starches from

**TABLE 5.4**

**Pasting properties of starches**

| Instrument used | Starch (%) | PV | BD | SV | Unit | PT (°C) | References |
|---|---|---|---|---|---|---|---|
| Rheometer | 16.6 | 4647–8303 | 2273–3344 | 1342–4722 | mPa. s | 72.4–73.9 | Sandhu and Siroha, 2017 |
| RVA | 10 | 2826 | 1313 | 2439 | mPa. s | 77.3 | Sharma et al., 2016 |
| BVA | 10 | 300 | 137 | 124.5 | BU | 73.8 | Shaikh et al., 2015 |
| RVA | 6 | 1665–1998 | 414–769 | 8–502 | cP | 88.1–90.2 | Khatkar et al., 2013 |
| RVA | 12 | 2708 | 1583 | 3229 | cP | 84.1 | Balasubramanian et al., 2014 |
| RVA | 12 | 202 | 107 | 20.8 | RVU | 75.2 | Choi et al., 2004 |
| RVA | 9 | 204–205 | 56–92 | 129–156 | RVU | 69.4–74 | Gaffa et al., 2004 |
| BVA | 6 | 100–180 | | | BU | 89.3–90 | Hoover et al., 1996 |
| BVA | 8 | 640 | | | BU | 70 | Wankhede et al., 1990 |

different cultivars varied from 63.4°C to 67.7°C, 69.3–71.6°C, and 74.5–76.3°C, respectively. Sharma et al. (2016) also studied gelatinization properties (To, Tp, Tc and ΔHgel) of pearl millet starch at 59.9°C, 62.6°C, 73.6°C and 11.9 J/g, respectively (Table 5.4). Khatkar et al. (2013) and Hoover et al. (1996) found approximate similar gelatinization temperatures for pearl millet starches. The difference in gelatinization properties between different studies is more likely due to the evaluating method, starch to water ratio, and the heating rate used for analysis. So, again, the data from different studies are impossible to compare.

## 5.10  DIGESTIBILITY PROPERTIES

The glycemic index (GI) concept is a tool for ranking foods with respect to their blood glucose raising potential (Kaur and Sandhu, 2010). GI is most appropriately used to compare foods within a category of foods. The digestibility of the human small intestine can be modified from a rapid digestion to indigestibility, which is the case in resistant starch. In general, digestible starches are hydrolyzed by the enzymes in the small intestine to yield free glucose that is then absorbed. Starch digestibility in the human digestive system can vary from rapid digestion to indigestibility (Lehmann and Robin, 2007). Starch digestion is an important metabolic response and the rate and extent of digestibility in the small intestine determine the eventual glucose level in the blood (Jenkins et al., 1982). Starch is classified into rapidly digestible starch (RDS), slowly digestible starch (SDS) and resistant starch (RS) on the basis of the rate and extent of its digestion. RDS is digested *in vitro* within 20 min, SDS is digested between 20 and 120 min, and RS is the starch that is not hydrolyzed after 120 min of incubation (Englyst et al.,

1992). RDS induces a rapid increase in blood glucose and insulin levels after ingestion. SDS prolongs the release of glucose, thus preventing hyperglycemia-related diseases (Mei et al., 2015). RS has been defined as the sum of starch and the product of starch degradation not absorbed in the small intestine, but is fermented in the large intestine of healthy individuals (Asp, 1992). Depending on the nature of inaccessibility, there are five classes of RS—RS1, RS2, RS3, RS4 and RS5 (Birt et al., 2013). RS1 is found in whole or coarsely ground grain where starch may be encased inside cells or in a strong protein matrix; RS2 consists of raw starch granules that resist amylase digestion, and RS3 is retrograded or recrystallized starch. RS4 is chemically modified starch with non-native chemical bonds formed, and RS5 is amylose complexed with lipid, usually a fatty acid (Zhao et al., 2011). Sandhu and Siroha (2017) reported RDS, SDS and RS content of pearl millet starch as 46.3–51.6%, 37.2–38.7% and 9.7–16.5%, respectively. Suma and Urooj (2015) observed RDS, SDS, RS content for pearl millet starches of 11.3–12.2%, 7.9–9.0% and 1.4–2.2%. Sharma et al. (2016) reported RS content for pearl millet starch at 2.4%. The size of starch granules may affect digestibility, as the relationship between surface area and starch volume, and, thus, contact between substrate and enzyme, decreases as the size of granule increases (Svihus et al., 2005). The health benefits of RS include prevention of colon cancer, hypoglycemic effects, substrate for growth of the probiotic microorganisms, reduction of gallstone formation, hypocholesterolemic effects and increased absorption of minerals (Sajilata et al., 2006).

## 5.11   STARCH MODIFICATION

Industrial use of native starches is limited, due to the instability of pastes and gels produced with them (Jyothi et al., 2005). Therefore, starch used in the food industry is often modified to overcome undesirable changes in product texture and appearance caused by retrogradation or breakdown of starch during processing and storage (Van-Hung and Morita, 2005). Therefore, starch is modified to improve its functional properties. Starch modification is generally done through derivatization such as etherification, esterification, cross-linking and grafting of starch or decomposition or physical treatment of starch using heat or moisture (Singh et al., 2007).

## 5.12   APPLICATIONS

Pearl millet starch is used in various industries. It could be concluded that incorporation of chemically modified pearl millet starches into custard samples resulted in the improvement of cold storage stability, textural and sensorial characteristics (Shaikh et al., 2017). Akande et al. (1991) observed that pearl millet starch showed comparable properties with corn starch for the formulation of tablets. Pearl millet starch is suitable for use as a binder and disintegrant in tablet formulations. Agbo et al. (2017) reported that pearl millet has been shown to be a good source of alpha-amylase that could be suitable for some biotechnological applications such as in starchy food processing industries, textile industries, brewing industries, etc. Odeku and Alabi (2007) suggested that native and modified pearl millet starch can be used as a disintegrate in tablets. Sharma et al. (2015) concluded that HMT pearl millet starch may find applications in heat processed food as well as in frozen food products.

## REFERENCES

Agbo, K. U., Eze, S. O., Okwuenu, P. C., Ezike, T. C., Ezugwu, A. L. and Chilaka, F. C. 2017. Extraction, purification and characterization of sprouting pearl millet alpha-amylase for biotechnological applications. *Journal of Plant Biochemistry and Physiology* 5: 1–6.

Akande, O. F., Deshpande, A. V. and Bangudu, A. B. 1991. An evaluation of starch obtained from pearl millet-*Pennisetum typhoides*. As a binder and disintegrant for compressed tablets. *Drug Development and Industrial Pharmacy* 17: 451–455.

Annor, G. A., Marcone, M., Bertoft, E. and Seetharaman, K. 2014a. Physical and molecular characterization of millet starches. *Cereal Chemistry* 91: 286–292.

Annor, G. A., Marcone, M., Bertoft, E. and Seetharaman, K. 2014b. Unit and internal chain profile of millet amylopectin. *Cereal Chemistry* 91: 29–34.

Asp, N. G. 1992. Resistant starch. Proceedings from the second plenary meetings of EURESTA. *European Journal of Clinical Nutrition* 46(Suppl. 2): S1.

Atwell, W. A., Hood, L. F., Lineback, D. R., Varriano-Marston, E. and Zobel, H. F. 1988. The terminology and methodology associated with basic starch phenomena. *Cereal Foods World* 33: 306–311.

Badi, S. M., Hoseney, R. C., & Finney, P. L. (1976). Pearl millet, 2: partial characterization of starch and use of millet flour in bread making. *Cereal Chemistry* 53, 718–72.

Bagley, E. B. and Christianson, D. D. 1982. Swelling capacity of starch and its relationship to suspension viscosity-effect of cooking time, temperature and concentration. *Journal of Texture Studies* 13: 115–126.

Balasubramanian, S., Sharma, R., Kaur, J. and Bhardwaj, N. 2014. Characterization of modified pearl millet (*Pennisetum typhoides*) starch. *Journal of Food Science and Technology* 51: 294–300.

Bao, J., Xiao, P., Hiratsuka, M., Sun, M. and Umemoto, T. 2009. Granule-bound SSIIa protein content and its relationship with amylopectin structure and gelatinization temperature of rice starch. *Starch/Stärke* 61: 431–437.

Beleia, A., Varriano-Marston, E. and Hoseney, R. C. 1980. Characterization of starch from pearl millets. *Cereal Chemistry* 54: 1096–1107.

Berski, W., Ptaszek, A., Ptaszek, P., Ziobro, R., Kowalski, G., Grzesik, M. and Achremowicz, B. 2011. Pasting and rheological properties of oat starch and its derivatives. *Carbohydrate Polymers* 83: 665–671.

Bertoft, E. 2004. Analysing starch structure. In A. C. Eliasson (Ed.), *Starch in Food-Structure, Function, and Application*. Cambridge: Woodhead Publishing, 57–71.

Bertoft, E., Piyachomkwan, K., Chatakanonda, P. and Sriroth, K. 2008. Internal unit chain composition in amylopectins. *Carbohydrate Polymers* 74: 527–543.

Biliaderis, C. G. 1998. Structures and phase transitions of starch polymers. In R. H. Walter (Ed.), *Polysaccharide Association Structures in Food*. New York: Marcel-Dekker, 57–168.

Birt, D. F., Boylston, T., Hendrich, S. et al. 2013. Resistant starch: promise for improving human health. *Advances in Nutrition* 4: 587–601.

Buleon, A., Colonna, P., Planchot, V. and Ball, S. 1998. Starch granules: Structure and biosynthesis. *International Journal of Biological Macromolecules* 23: 85–112.

Chiotelli, E. and Meste, M. L. 2002. Effect of small and large wheat starch granules on thermomechanical behaviour of starch. *Cereal Chemistry* 79: 286–293.

Choi, H., Kim, W. and Shin, M. 2004. Properties of Korean amaranth starch compared to waxy millet and waxy sorghum starches. *Starch/Stärke* 56: 469–477.

Chung, H. J., Liu, Q., Lee, L. and Wei, D. Z. 2011. Relationship between the structure, physicochemical properties and in vitro digestibility of rice starches with different amylose contents. *Food Hydrocolloids* 25: 968–975.

Cooke, D. and Gidley, M. J. 1992. Loss of crystalline and molecular order during starch gelatinisation: origin of the enthalpic transition. *Carbohydrate Research* 227: 103–112.

Copeland, L., Blazek, J., Salman, H. and Tang, M. C. 2009. Form and functionality of starch. *Food Hydrocolloids* 23: 1527–1534.

Da Silva, P. M. S., Oliveira, J. C. and Rao, M. A. 1997. The effect of granule size distribution on the rheological behavior of heated modified and unmodified maize starch dispersion. *Journal of Texture Studies* 28: 123–138.

Eliasson, A. C. and Gudmundsson, M. 1996. Starch: Physicochemical and functional aspects. In A. C. Eliasson (Ed.), *Carbohydrates in Food*. New York: Marcel Dekker, 431–503.

Englyst, H. N., Kingman, S. M. and Cummings, J. H. 1992. Classification and measurement of nutritionally important starch fractions. *European Journal of Clinical Nutrition* 46: S33–S50.

Estrada-León, R. J., Moo-Huchin, V. M., Ríos-Soberanis, C. R. et al. 2016. The effect of isolation method on properties of parota (*Enterolobium cyclocarpum*) starch. *Food Hydrocolloids* 57: 1–9.

Falade, K. O. and Okafor, C. A. 2013. Physicochemical properties of five cocoyam (*Colocasia esculenta* and *Xanthosoma sagittifolium*) starches. *Food Hydrocolloids* 30: 173–181.

Fannon, J. E., Hauber, R. J. and BeMiller, J. N. 1992. Surface pores of starch granules. *Cereal Chemistry* 69: 284–288.

Gaffa, T., Yoshimoto, Y., Hanashiro, I., Honda, O., Kawasaki, S. and Takeda, Y. 2004. Physicochemical properties and molecular structures of starches from millet (Pennisetum typhoides) and sorghum (*Sorghum bicolor* L. *Moench*) cultivars in Nigeria. *Cereal Chemistry* 81: 255–260.

Gani, A., Ashwar, B. A., Akhter, G., Shah, A., Wani, I. A. and Masoodi, F. A. 2017. Physico-chemical, structural, pasting and thermal properties of starches of fourteen Himalayan rice cultivars. *International Journal of Biological Macromolecules* 95: 1101–1107.

Gernat, C., Radosta, S., Damaschun, G. and Schierbaum, F. 1990. Supramolecular structure of legume starches revealed by X-ray scattering. *Starch/Stärke* 42: 175–178.

Ghiasi, K., Marston, V. K. and Hoseney, R. C. 1982. Gelatinization of wheat starch. II starche-surfactant interaction. *Cereal Chemistry* 59: 86–88.

Hoover, R. 2001. Composition, molecular structure, and physico-chemical properties of tuber and root starches: A review. *Carbohydrate Polymers* 45: 253–267.

Hoover, R., Hughes, T., Chung, H. J. and Liu, Q. 2010. Composition, molecular structure, properties, and modification of pulse starches: A review. *Food Research International* 43: 399–413.

Hoover, R. and Ratnayake, W. S. 2002. Starch characteristics of black bean, chick pea, lentil, navy bean and pinto bean cultivars grown in Canada. *Food Chemistry* 78: 489–498.

Hoover, R., Swamidas, G., Kok, L. S. and Vasanthan, T. 1996. Composition and physicochemical properties of starch from pearl millet grains. *Food Chemistry* 56: 355–367.

Hoover, R. and Vasanthan, T. 1994. The effect of annealing on the physico-chemical properties of wheat, oat, potato and lentil starches. *Journal of Food Biochemistry* 17: 303–325.

Jan, R., Saxena, D. C. and Singh, S. 2016. Pasting, thermal, morphological, rheological and structural characteristics of Chenopodium (*Chenopodium album*) starch. *LWT-Food Science and Technology* 66: 267–274.

Jane, J. and Chen, J. F. 1992. Effects of amylose molecular size and amylopectin branch chain length on paste properties of starch. *Cereal Chemistry* 69: 60–65.

Jane, J., Chen, Y. Y., Lee, L. F. et al. 1999. Effects of amylopectin branch chain length and amylose content on the gelatinization and pasting properties of starch. *Cereal Chemistry* 76: 629–637.

Jenkins, D. J. A., Thome, M. J., Camelon, K. et al. 1982. Effect of processing on digestibility and the blood glucose responses: a study of lentils. *American Journal of Clinical Nutrition* 36: 1093–1101.

Jenkins, P. 1994. *X-ray and Neutron Scattering Studies on Starch Granule Structure.* Ph.D. Thesis University of Cambridge.

Jyothi, A. N., Rajasekharan, K. N., Moorthy, S. N. and Sreekumar, J. 2005. Synthesis and characterization of low DS succinate derivatives of cassava (*Manihor esculenta crants*) starch. *Starch/Stärke* 57: 556–563.

Kaur, L., Singh, J., McCarthy, O. J. and Singh, H. 2007. Physicochemical, rheological and structural properties of fractionated potato starches. *Journal of Food Engineering* 82: 383–394.

Kaur, L., Singh, J., Singh, H. and McCarthy, O. J. 2008. Starch–cassia gum interactions: A microstructure-rheology study. *Food Chemistry* 111: 1–10.

Kaur, M. and Sandhu, K. S. 2010. In vitro digestibility, structural and functional properties of starch from pigeon pea (*Cajanus cajan*) cultivars grown in India. *Food Research International* 43: 263–268.

Kaur, M., Singh, N., Sandhu, K. S. and Guraya, H. S. 2004. Physicochemical, morphological, thermal and rheological properties of starches separated from kernels of some Indian mango cultivars (*Mangifera indica* L.). *Food Chemistry* 85: 131–140.

Kaur, M. and Singh, S. 2016. Physicochemical, morphological, pasting, and rheological properties of tamarind (*Tamarindus indica* L.) kernel starch. *International Journal of Food Properties* 19: 2432–2442.

Khatkar, S. K., Rajneesh, B. and Yadav, B. S. 2013. Physicochemical, functional, thermal and pasting properties of starches isolated from pearl millet cultivars. *International Food Research Journal* 20: 1555–1565.

Kim, Y. S., Wiesenborn, D. P., Lorenzen, J. H. and Berglund, P. 1996. Suitability of edible bean and potato starches for starch noodles. *Cereal Chemistry* 73: 302–308.

Krueger, B. R., Knutson, C. A., Inglett, G. E. and Walker, C. E. 1987. A differential scanning calorimetry study on the effect of annealing on gelatinization behaviour of corn starch. *Journal of Food Science* 52: 715–718.

Lee, M. H., Hettiarachchy, N. S., McNew, R. W. and Gnanasambandam, R. 1995. Physicochemical properties of calcium-fortified rice. *Cereal Chemistry* 72: 352–355.

Lehmann, U. and Robin, F. 2007. Slowly digestible starch-its structure and health implications: A review. *Trends in Food Science and Technology* 18: 346–355.

Lin, L. S., Guo, D. W., Zhao, L. X. et al. 2016. Comparative structure of starches from high-amylose maize inbred lines and their hybrids. *Food Hydrocolloids* 52: 19–28.

Lindeboom, N., Chang, P. R. and Tyler, R. T. 2004. Analytical, biochemical and physicochemical aspects of starch granule size, with emphasis on small granule starches: a review. *Starch-Stärke* 56: 89–99.

Mauro, D. J. 1996. An up to date of starch. *Cereal Food World* 41: 776–780.

Mei, J. Q., Zhou, D. N., Jin, Z. Y., Xu, X. M. and Chen, H. Q. 2015. Effects of citric acid esterification on digestibility, structural and physicochemical properties of cassava starch. *Food Chemistry* 187: 378–384.

Morris, E. R. 1989. Polysaccharide solution properties: origin, rheological characterization and implications for food system. In R. P. Millane, J. N. BeMiller, R. Cahndrasekavan (Eds.), *Frontiers in Carbohydrate Research-1: Food Applications*. London and New York: Elsevier Applied Science Publishers, 132–163.

Nep, E. I., Ngwuluka, N. C., Kemas, C. U. and Ochekpe, N. A. 2016. Rheological and structural properties of modified starches from the young shoots of Borassus aethiopium. *Food Hydrocolloids* 60: 265–270.

Odeku, O. A. and Alabi, C. O. 2007. Evaluation of native and modified forms of Pennisetum glaucum (millet) starch as disintegrant in chloroquine tablet formulations. *Journal of Drug Delivery Science and Technology* 17: 155–158.

Pérez, S., Baldwin, P. M. and Gallant, D. J. 2009. Structural features of starch granules I. In J. Bemiller, R. Whistler (Eds.), *Starch: Chemistry and Technology*. Vol 3 ed. Burlington: Academic Press, 5, 149–192.

Pérez, S. and Bertoft, E. 2010. The molecular structures of starch components and their contribution to the architecture of starch granules: A comprehensive review. *Starch* 62: 389–420.

Peterson, D. G. and Fulcher, R. G. 2001. Variation in Minnesota HRS wheats: starch granule size distribution. *Food Research International* 34: 357–363.

Punia, S., Siroha, A. K., Sandhu, K. S. and Kaur, M. 2019a. Rheological and pasting behavior of OSA modified mungbean starches and its utilization in cake formulation as fat replacer. *International Journal of Biological macromolecules* 128: 230–236.

Punia, S., Siroha, A. K., Sandhu, K. S. and Kaur, M. 2019b. Rheological behaviour of wheat starch and barley resistant starch (type IV) blends and their starch noodles making potential. *International Journal of Biological macromolecules* 130: 595–604.117.

Qiao, D. L., Xie, F. W. and Zhang, B. J. 2017. A further understanding of the multi-scale supramolecular structure and digestion rate of waxy starch. *Food Hydrocolloids* 65: 24–34.

Ratnayake, W. S., Hoover, R. and Warkentin, T. 2002. Pea starch: composition, structure and properties-a review. *Starch/Stärke* 54: 217–234.

Sajilata, M. G., Singhal, R. S. and Kulkarni, P. R. 2006. Resistant starch- A review. *Comprehensive Reviews in Food Science and Food Safety* 5: 1–17.

Salman, H., Blazek, J., Lopez-Rubio, A., Gilbert, E. P., Hanley, T. and Copeland, L. 2009. Structure-function relationships in A and B granules from wheat starches of similar amylose content. *Carbohydrate Polymers* 75: 420–427.

Sandhu, K. S. and Singh, N. 2005. Relationships between selected properties of starches from different corn lines. *International Journal of Food Properties* 8: 481–491.

Sandhu, K. S., Singh, N. and Kaur, M. 2004. Characteristics of the different corn types and their grain fractions: physicochemical, thermal, morphological, and rheological properties of starches. *Journal of Food Engineering* 64: 119–127.

Sandhu, K. S., Singh, N. and Lim, S. T. 2007. A comparison of native and acid thinned normal and waxy corn starches: Physicochemical, thermal, morphological and pasting properties. *LWT-Food Science and Technology* 40: 1527–1536.

Sandhu, K. S. and Siroha, A. K. 2017. Relationships between physicochemical, thermal, rheological and in vitro digestibility properties of starches from pearl millet cultivars. *LWT-Food Science and Technology* 83: 213–224.

Shaikh, M., Ali, T. M. and Hasnain, A. 2015. Post succinylation effects on morphological, functional and textural characteristics of acid-thinned pearl millet starches. *Journal of Cereal Science* 63: 57–63.

Shaikh, M., Ali, T. M. and Hasnain, A. 2017. Utilization of chemically modified pearl millet starches in preparation of custards with improved cold storage stability. *International Journal of Biological Macromolecules* 104: 360–366.

Sharma, M., Singh, A. K., Yadav, D. N., Arora, S. and Vishwakarma, R. K. 2016. Impact of octenyl succinylation on rheological, pasting, thermal and physicochemical properties of pearl millet (*Pennisetum typhoides*) starch. *LWT-Food Science and Technology* 73: 52–59.

Sharma, M., Yadav, D. N., Singh, A. K. and Tomar, S. K. 2015. Rheological and functional properties of heat moisture treated pearl millet starch. *Journal of Food Science and Technology* 52: 502–6510.

Singh, J., Kaur, L. and McCarthy, O. J. 2007. Factors influencing the physico-chemical, morphological, thermal and rheological properties of some chemically modified starches for food applications – A review. *Food Hydrocolloids* 21: 1–22.

Singh, J., McCarthy, O. J. and Singh, H. 2006. Physico-chemical and morphologi-cal characteristics of New Zealand Taewa (Maori potato) starches. *Carbohydrate Polymers* 64: 569–581.

Singh, N., Sandhu, K. S. and Kaur, M. 2004. Characterization of starches separated from Indian chickpea (*Cicer arietinum* L.) cultivars. *Journal of Food Engineering* 63: 441–449.

Singh, N., Singh, J., Kaur, L., Sodhi, N. S. and Gill, B. S. 2003. Morphological, thermal and rheological properties of starches from different botanical sources. *Food Chemistry* 81: 219–231.

Siroha, A. K., Sandhu, K. S. and Kaur, M. (2016). Physicochemical, functional and antioxidant properties of flour from pearl millet varieties grown in India. *Journal of Food Measurement and Characterization*, 10(2), 311–318.

Siroha, A. K. and Sandhu, K. S. 2018. Physicochemical, rheological, morphological, and in vitro digestibility properties of cross-linked starch from pearl millet cultivars. *International Journal of Food Properties* 21: 1371–1385.

Siroha, A. K., Sandhu, K. S., Kaur, M. and Kaur, V. 2019. Physicochemical, Rheological, Morphological and in Vitro Digestibility Properties of Pearl Millet Starch Modified at Varying Levels of Acetylation. *International journal of biological macromolecules* 131: 1077–1083.

Siroha, A. K., Sandhu, K. S. and Punia, S. 2019. Impact of octenyl succinic anhydride (OSA) on rheological properties of sorghum starch. *Quality Assurance and Safety of Crops & Foods* 131: 1077–1083.

Sowbhagya, C. M. and Bhattacharya, K. R. 1971. A simplified method for determination of amylose content in rice. *Starch/Starke* 23: 53–55.

Stanley D. W., Stone A. P. and Tung M. A. Mechanical Properties of Food. In *Handbook of Food Analysis*, Vol. I, Ch. 4. L. M. L. Nollet (Ed.), New York: Marcel Dekker, Inc., 93–136.

Suma P. F. and Urooj, A. 2015. Isolation and characterization of starch from pearl millet (*Pennisetum typhoidium*) flours. *International Journal of Food Properties* 18: 2675–2687.

Svegmark, K. and Hermansson, A. M. 1993. Microstructure and rheological properties of composites of potato starch granules and amylose: a comparison of observed and predicted structures. *Food Structure* 12: 181–193.

Svihus, B., Uhlenb, A. K. and Harstad, O. M. 2005. Effect of starch granule structure, associated components and processing on nutritive value of cereal starch: A review. *Animal Feed Science and Technology* 122: 303–320.

Tester, R. F. and Morrison, W. R. 1990. Swelling and gelatinization of cereal starches. *Cereal Chemistry* 67: 558–563.

Tester, R. F. and Morrison, W. R. 1992. Swelling and gelatinization of cereal starches. III. some properties of waxy and normal non waxy barley starches. *Cereal Chemistry* 69: 654–658.

Thitipraphunkul, K., Uttapap, D., Piyachomkwan, K. and Takeda, Y. 2003. A comparative study of edible canna (*Canna edulis*) starch from different cultivars. Part I. Chemical composition and physicochemical properties. *Carbohydrate Polymers* 53: 317–324.

Toker, O. S., Dogan, M., Canıyılmaz, E., Ersöz, N. B. and Kaya, Y. 2013. The effects of different gums and their interactions on the rheological properties of a dairy dessert: a mixture design approach. *Food and Bioprocess Technology* 6: 896–908.

Tsai, M. L., Li, C. F. and Lii, C. Y. 1997. Effects of granular structure on the pasting behaviour of starches. *Cereal Chemistry* 74: 750–757.

Van-Hung, P. and Morita, N. 2005. Effect of granule sizes on physicochemical properties of cross-linked and acetylated wheat starches. *Starch/Starke* 57: 413–420.

Wang, L., Xie, B., Xiong, G., Du, X., Qiao, Y. and Liao, L. 2012. Study on the granular characteristics of starches separated from Chinese rice cultivars. *Carbohydrate Polymers* 87: 1038–1044.

Wang, Y., Zhang, L., Li, X. and Gao, W. 2011. Physicochemical properties of starches from two different yam (*Dioscorea opposita Thunb.*) residues. *Brazilian Archives of Biology and Technology* 54: 243–251.

Wani, A. A., Singh, P., Shah, M. A., Wani, I. A., Götz, A., Schott, M. and Zacherl, C. 2013. Physico-chemical, thermal and rheological properties of starches isolated from newly released rice cultivars grown in Indian temperate climates. *LWT-Food Science and Technology* 53: 176–183.

Wani, I. A., Sogi, D. S., Hamdani, A. M., Gani, A., Bhat, N. A. and Shah, A. 2016. Isolation, composition, and physicochemical properties of starch from legumes: A review. *Starch/Stärke* 68: 834–845.

Wankhede, D. B., Rathi, S. S., Gunjal, B. B., Patil, H. B., Walde, S. G., Rodge, A. B. and Sawate, A. R. 1990. Studies on isolation and characterization of starch from pearl millet (*Pennisetum americanum* (L.) Leeke) grains. *Carbohydrate Polymers* 13: 17–28.

Waters, D. L., Henry, R. J., Reinke, R. F. and Fitzgerald, M. A. 2006. Gelatinization temperature of rice explained by polymorphisms in starch synthase. *Plant Biotechnology Journal* 4: 115–122.

Weaver, L., Yu, L. P. and Rollings, J. E. 1988. Weighted intrinsic viscosity relationships for polysaccharide mixtures in dilute aqueous solutions. *Journal of Applied Polymer Science* 35: 1631–1637.

Whistler, R. L., BeMiller, J. N. *Carbohydrate Chemistry for Food Scientist*. Amercian Association of Cereal Chemists: St. Paul, MN, 1997. 117–151.

Wu, Y., Lin, Q., Cui, T. and Xiao, H. 2014. Structural and physical properties of starches isolated from six varieties of millet grown in China. *International Journal of Food Properties* 17: 2344–2360.

Yoshimoto, Y., Egashira, T., Hanashiro, I., Ohinata, H., Takase, Y. and Takeda, Y. 2004. Molecular structure and some physicochemical properties of buckwheat starches. *Cereal Chemistry* 81: 515–520.

Yu, S., Ma, Y., Menager, L. and Sun, D. 2012. Physicochemical properties of starch and flour from different rice cultivars. *Food Bioprocess Technology* 5: 626–637.

Zhao, Y. S., Hasjim, J., Li, L., Jane, J. L., Hendrich, S. and Birt, D. F. 2011. Inhibition of azoxymethane-induced preneoplastic lesions in the rat colon by a cooked stearic acid complexed high-amylose corn starch. *Journal of Agricultural Food Chemistry* 59: 9700–9708.

Zhong, F., Li, Y., Ibáñez, A. M., Oh, M. H., McKenzie, K. S. and Shoemaker, C. 2009. The effect of rice variety and starch isolation method on the pasting and rheological properties of rice starch pastes. *Food Hydrocolloids* 23: 406–414.

Zhou, H., Wang, J., Zhao, H., Fang, X. and Sun, Y. 2010. Characterization of starches isolated from different Chinese Baizhi (*Angelica dahurica*) cultivars. *Starch/Stärke* 62: 198–204.

Zhu, F. 2014. Structure, physicochemical properties, and uses of millet starch. *Food Research International* 64: 200–211.

Zhu, F. 2016. Buckwheat starch: Structures, properties, and applications. *Trends in Food Science & Technology* 49: 121–135.

Zobel, H. F. 1988a. Starch crystal transformations and their industrial importance. *Starch/Stärke* 40: 1–7.

Zobel, H. F. 1988b. Molecules to granules-a comprehensive starch review. *Starch* 40: 44–50.

# 6 Impact of Different Modifications on Starch Properties

*Anil Kumar Siroha, Sneh Punia, Sukhvinder Singh Purewal, Loveleen Sharma and Ajay Singh*

## CONTENTS

## 6.1 INTRODUCTION

Starch is a major functional biopolymer, which can be utilized as a texturing agent in food and non-food applications. Starch contributes to the viscosity, texture, mouth-feel and consistency of food products (Blazek and Copeland, 2009). Foods made with native starches as an ingredient have low process tolerance for commercial manufacturing (Mason, 2009). During processing, the texture and appearance of the product is altered due to retrogradation, so, to overcome these undesirable changes, starch needs to be modified (Van Hung and Morita, 2005). Different methods are used to modify starch: physical, chemical and enzymatic (Table 6.1). Starch modification is mostly achieved through derivatization, for example, by esterification, etherification, cross-linking and decomposition (acid or enzymatic hydrolysis and oxidization of starch) or physical treatment such as heat and moisture (Sandhu et al., 2015).

The cross-linking agents commonly used to modify native starch are sodium trimetaphosphate, sodium tripolyphosphate, epichlorohydrin (EPI)

**TABLE 6.1**

**Modification of pearl millet starch**

| Modification | Major results and methodology | References |
|---|---|---|
| Acid hydrolysis | Degree of hydrolysis observed 80.5–88.2% (2.2 M HCl at 25°C for 20 days) | Hoover et al., 1996 |
| Acid modification | Swelling power, solubility, PV, BD, SB and FV decreased, PT increased (0.14 N HCL at 50°C for 8 h) | Balasubramanian et al., 2014 |
| HMT | Swelling power, solubility FV, SV increased, PV, TV, BV, PT decreased (starch (25–28% moisture content) heated for 3 h at 110°C) | Balasubramanian et al., 2014 |
| Enzymatic modification | Swelling power, solubility, PV, BD, SV, FV decreased and PT of modified starches was increased (Crude fungal amylase 0.1% was used. Starch–enzyme suspension was incubated at 37°C for 90 min in 0.04 M acetate buffer at pH 4.7) | Balasubramanian et al., 2014 |
| HMT | Swelling power, solubility, PV, BD increased while gelatinization temperatures (To, Tp, and Tc) and resistant starch content increase with increase in moisture content (20, 25 and 30%) as compared to native starch (starch with 20, 25 and 30% moisture content heated at 110±2°C for 8 h) | Sharma et al., 2015 |
| Acid thinning | Swelling power, PV, BV, SV, PT decreased and gelatinization parameters (To, Tp, Tc and ΔHgel) were increased when starch was modified with 1.0 M HCL concentration (0.1M and 1.0M HCL at 50°C for 2 h) | Shaikh et al. (2015) |
| Succinylation | Swelling power, solubility, PV, BD, SB increased and PT was decreased. Gelatinization temperatures (To, Tp, Tc) decreased and ΔHgel value was increased as compared to native starch (succinic anhydride (4%) pH 9.0–9.5 for 2 h) | Shaikh et al. (2015) |
| Succinylation | Swelling power, PV, cool paste viscosity, hot paste viscosity, BD increased while SB and PT were decreased. Gelatinization parameter (To, Tp, Tc and ΔHgel) were decreased. Yield stress and consistency index values were increased as compared to native starch. Granules showed slightly rough surfaces, some pores and cavities after modification (octenyl succinic anhydride (3%) pH 8 for 2, 3, 4, and 5 h). | Sharma et al., 2016 |

and phosphoryl chloride (Ratnayake and Jackson, 2008). Cross-linking of starch with EPI is the most common method used in the starch industry. Modification of starch with OSA (Octenyl succinic anhydride) is permitted at a maximum level of 3% based on the dry starch weight basis for use in food products in many countries (CFR, 2001). OSA starches stabilize the oil–water interface of an emulsion. The glucose part of the starch binds the water and lipophilic while the octenyl part binds the oil, thereby preventing separation of the oil and water phases (Murphy, 2000).

Chemical modification of starch by acetylation can be performed to significantly improve its physico-chemical and functional properties. Acetic anhydride in the presence of an alkaline agent is used to produce acetylated starches. The number of acetyl groups incorporated into the starch molecule during acetylation depends upon a number of factors, such as reactant concentration, starch source (Singh, Kaur et al., 2004), reaction time, and the presence of a catalyst (Agboola et al., 1991).

Current market trends are to produce more natural food components, so there is an increasing interest in improving the properties of native starches without using chemical modifications (Ortega-Ojeda and Eliasson, 2001). Heat–moisture treatment (HMT) is a physical modification technique applied to starches and it is considered safe as compared to chemical modifications. HMT is a physical modification method in which high temperatures (84–120°C) with intermediate moisture content (less than 35%) is applied to starches for varying time periods from 15 min to 16 h (Gunaratne and Hoover, 2002). HMT is influenced by the moisture content, heating temperature, heating time and types of starches. Annealing and HMT are the two common physical means by which the properties of starch are modified without rupturing the granule (Lim et al., 2001). Annealing refers to the treatment of starch in excess water (<65%, w/w) or at intermediate water contents (40–50%, w/w) at temperatures below the onset temperature of gelatinization. The physical aim of annealing is to approach the glass transition temperature, which enhances molecular mobility without triggering gelatinization (Hoover, 2010).

## 6.2 METHODS OF STARCH MODIFICATIONS

### 6.2.1 CHEMICAL MODIFICATION

Chemical modifications achieve the structural changes and incorporate new functional groups; these affect the physicochemical properties of the starches, making them fit for various industrial applications. Chemical modifications introduce functional groups into the starch molecule using derivatization reactions (e.g., etherification, esterification, cross-linking) or involve breakdown reactions (e.g., hydrolysis and oxidation) (Singh et al., 2007).

### 6.2.1.1 Modification with Cross-linking

The cross-linking method has been commonly used to modify native starch with various agents such as sodium trimetaphosphate (STMP), sodium tripolyphosphate (STPP), EPI and phosphoryl chloride (POCl$_3$) (Ratnayake and Jackson, 2008). Hirsch and Kokini (2002) studied the relative effects of different cross-linking agents on physical properties of starches and reported that POCl$_3$ has the ability to impart greater viscosity than STMP and EPI-treated granules. The cross-linking of starch was reported to be affected by various factors which include starch source, cross-linking reagent concentration and composition, the extent of substitution, pH, reaction time, and temperature (Lim and Seib, 1993; Chung et al., 2004;).

The common methods used for preparation of cross-linked starches were reported by various researchers (Wurtzburg, 1960; Zheng et al., 1999; Jyothi et al., 2006; Kaur et al., 2006). Kaur et al. (2006) reported that cross-linked starches prepared using lower $POCl_3$ concentration (1g/kg) showed higher swelling power (SP) and attributed it to the inclusion of a phosphate group inside the starch granules (Yoneya et al., 2003). The repulsion between the adjacent starch molecules caused by the negatively charged phosphate groups may have reduced the interchain associations and resulted in increased levels of hydrated swollen molecules, which resulted in increased SP (Lim and Seib, 1993). Siroha and Sandhu (2018) observed lesser solubility power for cross linked pearl millet starch. Jyothi et al. (2006) also reported that cross-linked starches exhibited lower solubility in water at 90°C than the native starch. The solubility decreased with an increase in reagent concentration, which could be attributed to the increase in the degree of cross-linking and is also consistent with the SP observed. Siroha and Sandhu (2018) found in their study that the SP of cross-linked starch was less than that of native starch. The SP of cross-linked starches decreased as compared to their native counterpart (Carmona-Garcia et al., 2009; Mirmoghtadaie et al., 2009; Ackar et al., 2010). Singh and Nath (2012) reported that the SP of cross-linked starches decreased at higher temperatures. This decrease may be due to formation of more gel mass that hinders the penetration of water into starch, reducing the swelling of cross-linked starch. SP and solubility were also decreased by modification, while paste clarity and freeze–thaw stability were influenced differently, due to a different extent of the chemical reaction between different starch varieties and EPI (Ackar et al., 2010). Koo et al. (2010) evaluated that cross-linking decreased the SP, solubility, and paste clarity of corn starch and the swelling factor was highly correlated with the degree of cross-linking. Kaur et al. (2006) also reported similar results for potato starches and observed that the decrease became greater as the reagent concentration increased. The starches treated with lower $POCl_3$ exhibited exceptionally higher SP than their counterpart native starches.

The degree of cross-linking (DC) of starches was determined using various methods described by Chatakanonda et al. (2000) and Kaur et al. (2006). The type and concentration of the reagent significantly affected the relative DC of the starches. The starches treated with a higher reagent concentration showed a higher degree of cross-linking, and vice versa. With a lower concentration of cross-linking reagent, DC was not observed (Kaur et al., 2006). Kurakake et al. (2009) reported that the degree of substitution of cross-linked starches increases with an increase in reaction time. Koo et al. (2010) reported optimal conditions for starch phosphorylation as reaction time of 4 h and reagent concentration of 1.5% (w/w). Siroha and Sandhu (2018) observed DC from 40.6–89.7% for pearl millet starches from different cultivars.

Cross-linking altered the pasting profile of the starches with peak viscosity (PV), breakdown viscosity (BV), final viscosity (FV), and setback viscosity

(SV) being reported to decrease as compared to their native counterparts (Kaur et al., 2006; Ackar et al., 2010; Rodrıguez-Marın et al., 2010; Siroha and Sandhu, 2018). The decrease in BV of the modified starches could be attributed to the formation of cross-links between starch molecules, which strengthens the swollen granules against breakage under conditions of high temperature and shear (Jyothi et al., 2006). Kurakake et al. (2009) reported that waxy corn starch shows a sharper pasting peak and its temperature was lower than normal corn starch. The major difference is the value of setback in the cooling process, with waxy starch having much less setback than normal starch. Siroha and Sandhu (2018) observed that PV, BV, trough viscosity (TV), SV, and FV of modified starches decreased, while pasting temperature (PT) was increased as compared to their native counterpart starches. Ackar et al. (2010) studied the influence of wheat variety and modification with EPI on starch properties. The study showed a maximum decrease in PV as well as BV and SV, and digestibility of starches can be reduced by proper selection of the EPI concentration used for modification.

Morphological characteristics of cross-linked starches have been studied by scanning electron microscopy (SEM). The effects of cross-linking were studied by various researchers (Kaur et al., 2006; Mirmoghtadaie et al., 2009; Koo et al., 2010; Siroha and Sandhu, 2018). Kaur et al. (2006) concluded that the cross-linking modification did not cause any detectable morphological change in potato starches. Similar results were reported by Mirmoghtadaie et al. (2009). Koo et al. (2010) reported that cross-linking caused slight changes in the structure of starch granules compared to native starch. Native starch granules were polygonal in shape with well-defined edges, whereas cross-linked starch granules exhibited a slightly rough surface and a black zone on the surface, whereas, Van Hung and Morita (2005) found that some granules were affected by exo-erosion due to chemical modification processes. The surfaces of the large granules exhibited greater damage after modification than those of the small granules. Siroha and Sandhu (2018), in their investigations, revealed that the cross-linking caused slight changes in the granular structure of starch as compared to pearl millet native starch.

Kaur et al. (2006) reported that starches treated with both the reagents (EPI and $POCl_3$) showed lower peak storage modulus (G′) and loss modulus (G″) than their native counterparts. The tendency of the starch pastes towards retrogradation increased considerably with increases in storage duration. However, the starches treated with 1g/kg $POCl_3$ exhibited much lower syneresis than the other cross-linked starches. Kim and Yoo (2010) reported cross-linking considerably reduced the SP, consistency index (K), apparent viscosity ($\eta$), and yield stress ($\sigma_o$) values of sweet potato starch, which significantly decreased with an increase in $POCl_3$ concentration. G′, G″, and complex viscosity ($\eta*$) of the cross-linked sweet potato starch pastes, determined using small deformation oscillatory rheometry, were higher than native starch, and these also decreased with an increase in $POCl_3$ concentration from 0.01% to 0.03%. tan δ value of starch gels shows that gels were elastic in nature. Siroha and Sandhu (2018) reported lesser G′ value for cross-linked starch than native pearl millet starch during heating and frequency sweep test. Yield stress and consistency index values were also decreased after the modification.

### 6.2.1.2 Starch Modification with Octenyl Succinic Anhydride (OSA)

The substitution of starch with OSA was first patented by Caldwell et al. (1953). In the present time, modification of starch with 3% OSA was permitted for use in food products in many countries based on the dry starch weight (CFR (Code of Federal Regulation), 2001). OSA starches stabilize the oil–water interface of an emulsion. The glucose part of starch binds the water and lipophilic, while the octenyl part binds the oil, so separation of the oil and water phases is prevented (Murphy, 2000). OSA modified starches are used in a variety of oil-in-water emulsions for food, pharmaceutical and industrial products, such as beverages and salad dressings, flavor encapsulating agents, clouding agents, processing aids, body powders, and lotions (Jeon et al., 1999; Park et al., 2004).

The degree of substitution (DS), defined as the average number of substitutions per anhydroglucose unit, varied between 0 and 3. The DS obtained did not go beyond the standard DS value of 3. The DS of OSA modified starches were evaluated using various methods. There are two types of titration methods for finding the DS of OSA starches. The first relies on the saponification of the product in an alkaline solution followed by titration of the excess alkali. As a general procedure, the OSA starch is suspended in a basic solution (commonly NaOH, (Bhosale and Singhal, 2006), or KOH, (Bao et al., 2003), which results in the saponification of the OS groups. The excess alkali present in the medium is then titrated with a hydrochloric acid solution with an indicator. In the second titration method (Hui et al., 2009), the OSA starch was dispersed in a hydrochloric acid–isopropanol solution. After filtration, the solid residue was washed with isopropanol until no $Cl^-$ was detected (using $AgNO_3$ solution) and redispersed in distilled water. This mixture was then boiled and the final solution was titrated using an indicator and an NaOH solution added until neutralization was achieved. For the same reason as explained above, the native unmodified starch was also titrated as a blank. A simplified method has also been suggested by Jeon et al. (1999) in which the product was simply dissolved in dimethyl sulphoxide and the solution was titrated against a 50 mM standard NaOH solution using phenolphthalein as the indicator.

The DS of starches depends on various factors, such as reactant concentration, reaction time, pH and the presence of a catalyst (Awokoya et al., 2011). Bhosale and Singhal (2006) studied the effect of the OSA–starch ratio, pH, temperature and time of the reaction on starch properties. The effects of these parameters were evaluated on the basis of the DS. The conclusion for amaranth-OSA starches was a reaction time of 6 h at 3% OSA–starch ratio at 30°C and pH 8.0 at 25% starch concentration. For waxy corn–OSA starch, all parameters were identical except for the reaction time of 24 h. The maximum DS achieved for both starches was 0.02, whereas Hui et al. (2009) reported the effect of the concentration of starch, reaction time, pH, and dilution with ethanol. They observed suitable parameters for the preparation of OSA starch from potato in aqueous slurry systems were concentration of starch slurry, 35% (in proportion to water, w/w), reaction period, 3 h, pH of the reaction system, 8.0, reaction temperature, 35°C, amount of OSA, 3% (in proportion to starch, w/w), OSA dilution-fold, five. The optimum reaction conditions of starch esterified with alkenyl succinic anhydride were pH 8.5–9.0, reaction temperature 23°C and 5% anhydride concentration (Jeon et al., 1999). Liu et al. (2008) reported maximal DS of OSA-modified waxy corn starch (0.0204) was predicted to occur when the starch concentration was 31.2%, pH was 8.6, the reaction temperature was 33.6° C, and the reaction time was 18.7 h. Segura-Campos et al. (2008) evaluated optimum treatment of a reaction with 3% OSA at pH 7 for 30 min, which produced 0.5083% succinyl groups and 0.00831 of substitution. Different starches show different DS on different reaction parameters. Huang et al. (2010) reported the reaction of corn starch with α-amylase when esterified with OSA. Results show that the α-amylase pre-treatment significantly decreased the DS of OS–starch compared with a non-pre-treated control group. Chen et al. (2014) studied physicochemical properties of corn starch and hydrothermally pre-treated OS–starch. Results showed that hydrothermal pre-treatments significantly increased the DS and reaction efficiency of H-OS-starch compared to control. The higher the pre-treatment temperature was, the deeper the OSA could penetrate into the internal starch granules.

Physicochemical properties of starches alter after the OSA modification. The SP of starches increased after the modification (Bhosale and Singhal, 2007; Arueya and Oyewale, 2015; Moin et al., 2016). SP is the capability of starch granules to hydrate at a particular temperature depending upon the availability of water (Lawal et al., 2011). Interactions between amorphous and crystalline constituents of starch granules could be assessed by determining SP and solubility (Ratnayake et al., 2002). Succinylated rice starches were found to have considerably higher SP in comparison to native rice starches. This observation is in accordance with earlier reports of an increase in SP after succinylation of yam (Lawal, 2012) and hybrid maize starches (Lawal, 2004). The increase in SP may be attributable to the weakening of the intermolecular hydrogen bond due to the introduction of a bulky OSA group (Perez et al., 1993). Solubility of the starches showed an increase after succinylation up to about 4% succinic anhydride. This may be attributed to structural reorganization, which tends to weaken the starch granules following succinylation (Arueya and Oyewale, 2015). No significant difference was observed in solubility after OSA modification for sorghum starches (Olayinka et al., 2011), who also studied the effect of pH on SP and solubility of modified starches. SP increased progressively from 2 to 12 and solubility reduced from pH 2 to 6, after which increase was observed in pH. It was observed that both native and modified starches had the highest SP and solubility at pH 12. The SP for all starches increased from pH 8 to 12 in the alkaline region, whereas little increase was observed in the acidic region (pH 2–6). Sandhu et al. (2015) claimed that amylose content of OSA modified potato starches decreased after the modification. OSA modification of starch carried out at different levels of pH also affected the amylose content of both the fractions, with modification under acidic conditions resulting in more lowering of amylose content. An increase in pH during modification, however, increased the amylose content for both the small and large granule fractions, with the highest amylose content being observed at pH 8. This may be due to the rapid reaction of amylose with OSA under more acidic conditions.

The pasting properties of starches also alter after the modification. Olayinka et al. (2011) reported OSA modification increased the PV of the succinylated starches and reduced SV and BV. There was no significant change in the PT. Arueya and Oyewale (2015) reported a decrease in pasting viscosities and the decrease did not follow the order of substitution. Sandhu et al. (2015) reported that the variations in pH during modification had great effects on the pasting characteristics of both starch fractions (small and large) and PV was decreased at pH 4 and pH 6, as compared to their native counterpart. Chung et al. (2010) observed that the pasting viscosities of corn starches treated under acidic condition were lower as compared with OSA starch treated under basic pH conditions. The pH for OSA substitution had great effects on the pasting characteristics during dry heating. When the OSA starch reacted at pH 4 and pH 6 with dry heating at 130°C for 2 h, the PV, BV, SV and FV substantially decreased in comparison to their unheated OSA starches, and the greatest decrease in pasting viscosity was observed under the acidic condition of pH 4. Song et al. (2010) reported the process of OSA modification of starch derived properties which had higher viscosities, better paste clarity and freeze–thaw stability than the native starch. Texture results showed that the hardness, springiness, cohesiveness, and chewiness of a sausage increased as OSA was added. Shogren et al. (2000) suggested that the OS group facilitated an increase in water percolation within the granules as a result of steric hindrance and repulsion, allowing an increase in swelling volume. Thirathumthavorn and Charoenrein (2006) claimed that the network formation of amylose–OSA inclusion complexes could enhance the PV. Moin et al. (2016) and Chung et al. (2010) found that the PT of native starch was higher than that of modified starches, suggesting ease of gelatinization in succinylated starch granules. The PT further reduced with the increase in the succinic anhydride level, indicating weakening of the granules, whereas, Arueya and Oyewale (2015), in their study, found that the PT was increased after modification. Amylose is known to limit the expansion of starch paste during heating, leading to an increase in the PT of starch. This conclusion is valid, as the amylose content obtained increased after succinylation. The ability of starch to imbibe water and swell is primarily dependent on the PT. The higher the PT, the faster is the tendency for the paste to be formed.

OSA modified starches show different morphological properties as compared to native starches. SEM investigations showed that the dry heated OSA modification caused drastic changes on the surface of starch granules, with modified large granule fraction (LGF) potato starch exhibiting rough surfaces with lost definition of its edges. Comparison of both modified small granule fraction (SGF) and LGF potato starches revealed that the modification had a more pronounced effect on SGF potato starch when compared to their counterpart modified LGF starch. The granules of OSA modified SGF potato starch were also ruptured (Sandhu et al., 2015). While Arueya and Oyewale (2015) concluded that no obvious differences were observed in the morphology of the native and succinylated starch, Moin et al. (2016) observed deformation of starch granular structure after modification, which could be due to strong alkaline conditions used during starch modification. Deformation of starch granules is also reported in yam starch after succinylation (Lawal, 2012). It could be observed that the extent of deformation and aggregation of modified starch granules increased with the level of succinic anhydride added. Succinylation altered the granular structure of rice starch, as rough edges, cavities and fragments could be observed in modified starch micrographs. It is evident from the figures that fusion of starch granules took place after the addition of succinic anhydride. Roughness on the surface of granules promotes clump formation in starch. Similar effects in the morphology of granules were also observed due to derivitization of rice starch with an acetyl group (Gonzalez and Perez, 2002).

Park et al. (2004) studied rheological properties of OSA modified corn starch pastes (5%, w/w), at different OSA contents (0, 1.0, 1.5, 2.0 and 2.5%, w/w) and evaluated their steady and dynamic shear properties. The OSA starch pastes had high shear-thinning behaviors and their flow behaviors were described by power law, Casson, and Herschel-Bulkley models. The magnitudes of K and $\sigma_O$ increased with the increase in OSA content and the decrease in temperature. A dynamic frequency sweep test showed that both $G'$ and $G''$ of OSA starch pastes increased with the increase in OSA content.

Hadnadev et al. (2015) reported functionality of OSA starch-stabilized emulsions as fat replacers in cookies. The effects of the incorporation of structured oil (in the form of 50% and 70% oil-in-water emulsions) instead of unstructured oil (50% and 70%) or traditional shortening in cookie formation was studied. It was determined that the replacement of vegetable fat with emulsions and unstructured oil affected the decrease in dough elastic modulus and increase in cookie firmness. Cookies containing oil in the form of emulsion expressed higher dough strength and lower cookie spread in comparison to those containing unsaturated oil. Although all the cookies were sensorily acceptable, the one containing traditional shortening had superior sensory characteristics.

### 6.2.1.3   Modification with Acetic Anhydride

Chemical modification involving acetylation is a widely used method for starch modification to improve functional or physicochemical properties and increase their application in a variety of food products. Acetylated starch is prepared commonly with acetic anhydride and vinyl acetate. According to The Joint FAO/WHO Expert Committee on Food Additives (JECFA), starch acetate is basically defined as a starch ester prepared by the addition of acetic anhydride or vinyl acetate to starch under alkaline conditions. Addition of these esterification reagents results in the replacement of hydroxyl groups with acetyl moieties. However, according to the Food and Drug Administration, the starch acetate prepared by any of the methods should not have more than 2.5% acetyl groups (Codex, 1996; JECFA, 2001).

The DS indicates the amount of substitute introduced in the chemically modified starch during the reaction. In acetylated starch, part of the hydroxyl groups in anhydroglucose units has been converted to acetyl groups (Wani et al., 2012). Singh et al. (2012) found that DS and percentage acetylation increased from 0.050 to 0.081 and 1.31% to 2.10%, respectively, with the increase in the level of addition of acetic anhydride from 1.25 to 6.25 g/100 g. Acetyl content between 4.68% and 5.97% and 3.43% to 4.68%, respectively, for acetylated potato and corn starches, using 2–12% w/w acetic anhydride has been reported by Singh, Chawla et al. (2004). Acetyl content of acetylated starches from other sources (potato, waxy corn, corn, wheat, field pea, and lentil) has been reported between 1.01% and 2.80% using 5% and 10% of acetic anhydride (Vasanthan et al., 1995). These differences in acetyl content may be due to the use of varied concentrations of reactants and the difference in starch structure from their botanical sources. Acetic anhydride in the presence of an alkaline agent, such as sodium hydroxide, is used to produce acetylated starches (Wurzburg, 1978). The number of acetyl groups incorporated into the starch molecule during acetylation depends upon a number of factors, such as reactant concentration, starch source (Singh, Chawla et al., 2004), reaction time, and the presence of a catalyst (Agboola et al., 1991).

Physicochemical properties of acetylated starches differ from those of their native counterpart. Many researchers have reported on the physicochemical properties of starches (Shon and Yoo, 2006; Singh et al., 2012; Wani et al., 2015; Sun et al., 2016; Siroha et al., 2019). The SP and solubility of acetylated starch increased after acetylation (Shon and Yoo, 2006; Simsek et al., 2012). The changes in the SP and solubility upon acetylation may be attributed to the introduction of hydrophilic substituting groups that retain water molecules to form hydrogen bonds in the starch granules (Betancur et al., 1997). Moreover, the access of water molecules to the amorphous regions could have also been increased by the introduction of acetyl groups. A similar increase in the SP and solubility upon acetylation has been reported in earlier investigations (Liu et al., 1997; Singh, Chawla et al., 2004; Adebowale et al., 2006). The introduction of acetyl groups reduces the bond strength between starch molecules and, thereby, increases the SP and solubility of the starch granule, decreasing the retrogradation of the starch and improving the clarity and freeze–thaw stability (Chi et al., 2007). The amylose content of acetylated starches increased as compared to their native counterpart (Betancur et al., 1997; Singh, Chawla et al., 2004, 2012). However, some researchers reported that amylose content of acetylated starches decreased as compared to native starches (Wani et al., 2012, 2015). The differences may be due to the introduction of acetyl groups to starch chains, which impeded the formation of the helical structure of amylose in some areas through steric hindrance and, in consequence, the formation of amylose iodine complex, so that actual amylose content was underestimated (Gonzalez and Perez, 2002).

Pasting behavior of starch was significantly affected by acetylation. Pasting properties are commonly used to study the effect of heating or cooling cycles on the behavior of native and modified starches. Singh et al. (2012) reported that PV, HPV, BV and CPV were decreased after the acetylation. Simsek et al. (2012) and Lawal et al. (2015) also concluded that, after the acetylation, the PV was decreased. Sun et al. (2016) found that pasting parameters, such as the PV, BV, FV, SV and PT were lower for acetylated starch than for native starch. There are two factors that may account for the low viscosity of the acetylated starches. First, the introduction of acetyl groups weakened and disintegrated the ordered structure of the native starch during the modification process (Saartrat et al., 2005). Second, the acetylation process resulted in depolymerization of the starch samples. The depolymerization resulted in a reduction of the molecular weight and, thus, caused the viscosity to decrease (Sun et al., 2016). Shubeena et al. (2015) observed that acetylated starches show higher PV, BV, TV, SV and FV as compared to their native counterpart. Wani et al. (2012) also observed that the PV of acetylated starches was higher than native starches. Higher PV in acetylated starches than in respective native starches was probably due to a decrease in the extent of molecular association within amorphous regions of acetylated starch granules, causing higher swelling and extensive amylose leaching during the heating period (Hoover and Hadziyev, 1981).

Different researchers have reported the morphological properties of acetylated starches. Singh et al. (2012) found that, after acetylation, change occurred in the external morphology leading to surface roughness, formation of cavities, and fusion of few starch granules was observed. The size of granules was also observed to slightly decrease with the acetylation treatment. These changes may be attributed to the gelatinization of the surface of starch granules with NaOH during the acetylation treatment. Lawal et al. (2015) reported that acetylated forms of the three botanical starches (corn, cassava, and sweet potato) maintained a similar particle shape to their native counterparts but exhibited increased granular aggregation. The slight modification of the morphology of starch acetates may be attributed to their low DS of less than 0.2. Singh, Kaur et al. (2004) concluded that the granule size distributions of acetylated starches were slightly different from those from the native starches. Among the potato starches, the granule size distributions showed a very small increase from the Kufri Chandermukhi and Kufri Sindhuri cultivars; however, no significant difference in granule size distributions of starches from other cultivars and corn starch was observed. It may also be inferred that the granule size distributions of acetylated starches may not be accurately measured due to fusion of the small granules. The granular structure of native corn and potato starches showed significant variation in size and shape when viewed by SEM. Shubeena et al. (2015) evaluated that morphology of the starch granules appears to be altered by the modification reaction. It seems that starch granules agglomerate upon acetylation. According to Singh, Kaur et al. (2004), the introduction of acetyl groups to the starch molecules results in an increase in intermolecular hydrogen bonding, which leads to the fusion of starch granules. Acetylation does not affect the morphology of starch granules (Sodhi and Singh, 2005; Colussi et al., 2014; Bartz et al., 2015).

Fornal et al. (2012) reported that greater changes in granule appearance occurred in larger granules than in smaller ones. Changes in granule size during acetylation was the main reason that the acetylated starch differed in granularity from the native starch, but still the granularity of the preparation depended on the applied reagent (vinyl acetate or acetic acid anhydride) (Garcia et al., 2012) and on conditions (time) of the reaction (Mbougueng et al., 2012).

Colussi et al. (2014) reported that acetylated rice starches showed a decrease in the intensities of the peaks compared to the native ones, with the exception of low amylose starch acetylated for 90 min of reaction. Acetylation reduced the crystallinity of rice starches, and the lowest values of relative crystallinity were observed in acetylated starches with the highest DS. Sha et al. (2012) reported that, with the increase in the proportion in acetyl content of the rice starch, crystallinity became gradually lowered and the diffraction peak also reduced in turn. They described that the changes in the diffraction patterns indicated that the intermolecular hydrogen bonding interaction was damaged. Mbougueng et al. (2012) reported the same XRD trend for native potato starch and the acetylated ones. However, the crystalline index shows an increase in crystalline region of the starches with acetylation. Singh et al. (2012) also observed that, after the acetylation, diffraction patterns did not change.

Shon and Yoo (2006) reported that acetylated rice starch pastes at 25°C had high shear-thinning behavior (n = 0.17–0.20) with high Casson yield stress ($\sigma oc$ = 47.4–84.1 Pa). Magnitudes of K and $\sigma oc$ increased with an increase in acetyl substitution degree. Magnitudes of G′, G″ and complex viscosity ($\eta*$) of acetylated starch pastes increased with an increase in the degree of acetyl substitution. Singh, Kaur et al. (2004) reported that acetylated starches showed higher peak G′, G″ and lower tan δ than their counterpart native starches during heating.

### 6.2.1.4   Modification of Starch with Acid

Acid hydrolysis is an important chemical modification that can significantly change the structural and functional properties of starch without disrupting its granular morphology. A deep understanding of the effect of acid hydrolysis on starch structure and functionality is of great importance for scientific research into starch and its industrial applications (Wang and Copeland, 2015). The hydrolysis rate was not constant during the course of acid modification among different starches. Kerr (1952) demonstrated that in early stages of acid modification, the amount of amylose or linear fraction in starch increased, suggesting that acid preferentially hydrolyzed amylopectin. Kang et al. (1994) reported that the amylose and amylopectin of rice starch were hydrolyzed simultaneously in the first phase of acid degradation, but amylose was more affected. In acid modification, the hydroxonium ion attacks the glycosidic oxygen atom and hydrolyses the glycosidic linkage. Acid modification changes the physicochemical properties of starch without destroying its granule structure. In addition, the physicochemical properties of acid-thinned starches differ according to their origin and the conditions of preparation (Shi and Seib, 1992).

Lawal et al. (2005) reported that acid thinning reduced the swelling power and solubility of the native starch. The increase in crystallinity is responsible for a reduction in swelling capacity of the acid-thinned starch, since swelling is restricted by the stiffness of the entangled amylopectin network in the crystalline region of the starch (Cairns et al., 1990). Abdorreza et al. (2012) investigated whether swelling power decreased and solubility increased by increasing the duration of acid treatment. It is likely that acid treatment increases the fraction of low molecular weight chains with hydroxyl groups and, therefore, improves solubilization in water. A study conducted by Balasubramanian et al. (2014) on acid hydrolysis of pearl millet reported that, after this chemical modification, the swelling and solubility of starch was reduced. Shaikh et al. (2015) reported that swelling power was increased when the starch was modified at low concentration (0.1 M), while at higher concentrations it decreased. A decrease in solubility was observed for acid-treated pearl millet starch.

A significant reduction in the values of pasting parameters (PV, TV, BV, FV and SV) was observed with acid thinning but the reduction was more pronounced in the case of waxy corn starches (Sandhu et al., 2007). Singh et al. (2009) studied the effect of the duration of acid treatment on sorghum starch. Pasting parameters showed a decreasing trend upon increase in the duration of acid hydrolysis. The reduction in PV through acid treatment may be attributed to an increase in the hydrolysis of amorphous regions and production of low molecular weight dextrins. Thirathumthavorn and Charoenrein (2005) reported that the setback values of acid-treated rice starches were lower than those of native rice starch. Shaikh et al. (2015) stated that after acid treatment time to reach PV was significantly reduced, PV, BV and SV was decreased for acid-thinned pearl millet starch. Balasubramanian et al. (2014) also reported that acid-modified starch had lower breakdown and higher setback viscosity as compared to native starch.

Wang and Wang (2001) studied the physicochemical properties of acid-thinned corn, potato and rice starches. They found that acid modification changed the physicochemical properties of starches without destroying their granule structure, and the properties of acid-thinned starches differ according to their origins. Sandhu et al. (2007) reported that acid thinning does not affect the granule structure of starch. This may be due to the low level of hydrolysis. Similar results were observed by Singh et al. (2009). Shaikh et al. (2015) observed granular destruction with increased dents and indentations in pearl millet starch after acid thinning at a high level of acid concentration. The acid acts by first attacking the surface and forming cracks there at the first hydrolysis step. Acids cause surface alterations and degrade the external part of the granule by exo-corrosion. When endo-corrosion occurs, the internal part of the granule is corroded through small cracks through which acids could penetrate the granule (Jiping et al., 2007).

Acid-thinning treatment preferentially attacks the amorphous regions in the granules; the crystallites are decoupled from, and no longer destabilized by, the amorphous parts. Consequently, the starch crystallites of the acid-thinned starches melt at higher temperatures (Garcia et al., 1996; Jenkins and Donald, 1997; Hoover, 2000). Singh et al. (2009) reported that onset temperature decreased and end temperature increased acid hydrolyzed starches. This increase could be due to longer amylopectin double helices resulting from acid hydrolysis than in unhydrolyzed amylopectin molecules, because branch points in unhydrolyzed amylopectin may reduce the length of helix-forming side chains (Morrison et al., 1993). Shaikh et al. (2015) observed that onset, peak and end temperatures decreased when low concentrations were used for acid hydrolysis and the reverse was observed for high concentrations. In both cases, enthalpy of the gelatinization value increased for pearl millet starches.

## 6.2.2 Physical Modification

### 6.2.2.1 Starch Modification with Heat Moisture Treatment

The physical modification of starch by moisture, heat, shear, or radiation has been gaining wider acceptance because no by-products of chemical reagents are present in the modified starch. Chemical modification is widely implemented, but there is also a growing interest in the physical modification of starch, especially in food applications (Zavareze and Dias, 2011). A major advantage of physical modification is that starch is considered to be a natural material and a highly safe ingredient, so its presence and quantity in food is not limited by legislation (Bemiller, 1997). The term HMT is used when low moisture levels (<35% w/w) are applied, whereas, annealing refers to treatment of starch in excess water (<65% w/w) or at intermediate water (40–55% w/w). Both HMT and annealing occur below the onset temperature of gelatinization and have been shown to modify the structural arrangement of starch chains to different degrees (Hoover, 2010). Different methods described by various researchers are used to modify the starches with HMT (Hormdok and Noomhorm, 2007; Juansang et al., 2015; Lee and Moon, 2015; Van Hung et al., 2016).

Several researchers have studied the effect of HMT on the SP of various starches (Klein et al., 2013; Sharma et al., 2015; Sui et al., 2015). In all of these studies, the authors observed a reduction in the SP of HMT starches. This reduction in SP has also been attributed to increased crystallinity, reduced hydration (Waduge et al., 2006), increased interactions between amylose and amylopectin molecules, strengthened intramolecular bonds (Jacobs et al., 1995), the formation of amylose–lipid complexes (Waduge et al., 2006) and changes in the arrangements of the crystalline regions of starch (Hoover and Vasanthan, 1994b). Starch solubility results from the leaching of amylose, which dissociates from, and diffuses out of, granules during swelling. This leaching represents a transition from order to disorder within the starch granules that occurs when starch is heated with water (Tester and Morrison, 1990). Olayinka et al. (2008) noted a decrease in the solubility of hydrothermally treated white sorghum starch compared to native starch. This solubility decrease was enhanced by increasing the moisture content of the treated starches from 18% to 27%. Sharma et al. (2015) concluded that starch solubility was decreased after the modification and it was affected by the amount of leached amylose during hydration and swelling. The formation of cross-linkages in the amorphous region and, thus, an increase in crystallinity during heat moisture treatment might be a possible reason for a decrease in SP and solubility. Van Hung et al. (2016) reported that the degree of solubility of the native normal and high-amylose rice starches were less than 10%, even at heating temperatures up to 90°C, whereas the solubility of the native waxy rice starch was slightly higher when heating was done at 80–90°C. Starches treated with citric acid and lactic acid had significantly higher degrees of solubility, especially in hot water, as compared to the native starches. Zhang et al. (2010) reported that amylose content, SP, solubility, as well as WAC and OAC of native starch, were higher than in treated canna edulis ker starches.

HMT promotes intense changes in starches; it alters the pasting profile of starches significantly. Sharma et al. (2015) concluded that the PT increased with an increase in moisture content during HMT. Other pasting parameters, that is, PV, BV, cool paste viscosity (CPV) and SV, decreased significantly (p <0.05) as the moisture level of the samples increased after HMT. The increase in the PT may be attributed to a rearrangement of the amorphous region into a more compact crystalline region, as well as the formation of cross-linkages. Therefore, more heat is required for structural disintegration and paste formation. Klein et al. (2013) reported a reduction in the PV of all starch sources with the increase in HMT temperature and with the application of dual HMT compared to the respective native starches, except when dual HMT at 120°C was done for rice and cassava starches. Sui et al. (2015), in their study, concluded that HMT starches exhibited significantly lower PV, FV, BV and SV than native starches. The pasting properties of HMT products were affected to different degrees by various reaction conditions. When the length of heating was constant, the extent of the decrease in PV and FV of normal maize starch (NMS) as a function of moisture content during HMT was more pronounced than that of waxy maize starch (WMS). Olayinka et al. (2008) evaluated that PV, BV and SV was decreased after the HMT treatment. Singh et al. (2011) found that after the HMT pasting properties, PV, SV, BV and FV reduced as compared to their native counterpart except for starch treated at 20% moisture content, which shows an increase in BV and SV.

The effect of HMT on granule morphology has been shown by various researchers. Sharma et al. (2015) showed that HMT induced some dents/holes on the surface of starch granules, possibly due to the rearrangement of the molecular structure, as well as disintegration. The severity increased with an increase in moisture content during HMT. Loss of physical integrity with degradation of the granular surface (characteristic of partial gelatinization) was also observed in HMT samples. Lee et al. (2012) suggested that this may be due to partial swelling and disruption of starch granules in the presence of abundant water. Singh et al. (2011) observed, using SEM, that HMT causes a considerable change in granule morphology and the granule size of HMT starch was lower than native counterpart starch. However, the reduction in size was prominent in HMT40 (HMT at 40% moisture content) starches. Annealed (ANN) starch showed the presence of granules with roughened surface and grooves. HMT40 starch showed the presence of granules which exhibited a structure like that of bread dough. The results indicated that the effect of heat–moisture treatment on the morphology of starch granules was moisture dependent. Wheat, oat, lentil and potato starches have been reported to show no change in granule morphologies upon annealing (Hoover and Vasanthan, 1994a). Lee et al. (2012) concluded that HMT did not alter the shape or size of the waxy potato starch granules, except for starch treated for 5 h at 150°C and 30% moisture. However, the surfaces of all HMT samples showed signs of cracks and dents and the cross-section of each starch granule displayed a large hollow central region. These results may have been caused by partial swelling and disruption of starch granules by abundant water molecules. The hilum, usually central, is the initiation and growth point of a starch granule; this central tissue is believed to be less organized than other granule tissues (Fannon et al., 2004). Watcharatewinkul et al. (2009) reported that HMT did not alter the shape or size of starch granules. Jiranuntakul et al. (2011) evaluated HMT starches from normal rice, waxy rice, normal corn, waxy corn, normal potato and waxy potato, and found that the HMT did not change the size, shape, and surface characteristics of corn and potato starch granules, while surface change/partial gelatinization was found on the granules of rice starches.

Different studies were carried out by different scientists with the aim of improving the properties of starches through combining different treatments. Van Hung et al. (2016) reported that acid and HMT reduced SP and viscosity, but increased solubility of the starches, while the crystalline structure did not change. Klein et al. (2013) studied the effects of single and dual HMT of rice, cassava and pinhao starches at 100°C and 120°C. The temperature variation affected the starch properties more than the single or dual HMT. The starch subjected to single HMT at 120°C was the most applicable to food applications, where low SP, low viscosity and high thermal stability are necessary. Juansang et al. (2015) evaluated the effect of plasticizers (propanol, propylene glycol, glycerol, erythritol, xylitol and sorbitol) on canna starch during HMT. The least change in paste viscosity was found when water was used as a plasticizer. The viscosity of the modified starches decreased as the molecular weight of plasticizers decreased. Plasticizer content in starch granules increased with decreasing molecular weight of the plasticizer, as well as with increased soaking time.

Klein et al. (2013) and Sui et al. (2015) observed that HMT starches did not alter the inherent diffraction pattern of native starches. Hoover and Vasanthan (1994b) and Gunaratne and Hoover (2002) found that the X-ray pattern of B-type tuber starches could be altered to A-type starches with HMT, whereas the X-ray pattern of A-type cereal starches remained unchanged. Changes in the X-ray intensities in cereal starches (A-type) during HMT have been suggested to be a result of moisture and thermal energy causing a strengthening of the association of the adjacent double helices and, consequently, an increased amount of direct hydrogen bonds (Zavareze and Dias, 2011). Pinto et al. (2012) concluded that, compared to native starch, there was an increase in peak intensity for all HMT starches, except at 25% moisture content and 120°C.

## 6.2.2.2 Annealing (ANN)

ANN is the hydrothermal treatment of starch in the presence of excess water for an extended period of time. This treatment is performed at a temperature above the glass transition but below the gelatinization temperature of starch (Gomes et al., 2005; Jayakody and Hoover, 2008). HMT is carried out under restricted moisture content (10–30%) and higher temperatures (90–120°C), while ANN involves a large excess of water (50–60%) and relatively low temperatures that fall below the gelatinization point (Maache-Rezzoug et al., 2008). ANN is a physical reorganization of starch granules, which promotes both an increase in the gelatinization temperature and enthalpy and a decrease in the temperature range at which this endothermic phenomenon occurs. The heat required for gelatinization of starch is inversely proportional to the area of the starch crystalline region; thus, it is a technique with greater viability for starches with larger amorphous regions (Tester and Debon, 2000). The process of annealing is associated with the reorganization of starch amylose and amylopectin double helices to result in higher crystallinity, but this slightly affects the crystalline and molecular order (Zavareze and Dias, 2011).

## REFERENCES

Abdorreza, M. N., Robal, M., Cheng, L. H., Tajul, A. Y., & Karim, A. A. 2012. Physicochemical, thermal, and rheological properties of acid-hydrolyzed sago (*Metroxylon sagu*) starch. *LWT-Food Science and Technology* 46: 135–141.

Ackar, D., Babic, J., Subaric, D., Kopjar, M., & Milicevic, B. 2010. Isolation of starch from two wheat varieties and their modification with epichlorohydrin. *Carbohydrate Polymers* 81: 76–82.

Adebowale, K. O., Afolabi, T. A., & Olu-Owolabi, B. I. 2006. Functional, physicochemical and retrogradation properties of sword bean (*Canavalia gladiata*) acetylated and oxidised starches. *Carbohyrate Polymers* 65: 93–101.

Agboola, S. O., Akingbala, J. O., & Oguntimein, G. B. 1991. Production of low substituted cassava starch acetates and citrates. *Starch/Stärke* 43: 13–15.

Arueya, G. L., & Oyewale, T. M. 2015. Effect of varying degrees of succinylation on the functional and morphological properties of starch from acha (*Digitaria exilis Kippis Stapf*). *Food Chemistry* 177: 258–266.

Awokoya, K. N., Nwokocha, L. M., Moronkola, B. A., & Moronkola, D. O. 2011. Studies on the isolation, structural and functional properties of starch succinate of cocoyam (*Colocasia antiquorum*) pelagia research library. *Der Chemica Sinica* 2: 228–244.

Balasubramanian, S., Sharma, R., Kaur, J., & Bhardwaj, N. 2014. Characterization of modified pearl millet (*Pennisetum typhoides*) starch. *Journal of Food Science and Technology* 51: 294–300.

Bao, J., Xing, J., Phillips, D. L., & Corke, H. 2003. Physical properties of octenyl succinic anhydride modified rice, wheat, and potato starches. *Journal of Agricultural and Food Chemistry* 51: 2283–2287.

Bartz, J., Goebel, J. T., Giovanaz, M. A., Zavareze Elessandra, D. R., Schirmer, M. A., & Dias, R. G. 2015. Acetylation of barnyard grass starch with acetic anhydride under iodine catalysis. *Food Chemistry* 178: 236–242.

Bemiller, J. N. 1997. Starch modification: Challenges and prospects. *Starch/Stärke* 49: 127–131.

Betancur, A. D., Chel, G. L., & Canizares, H. E. 1997. Acetylation and characterization of Canavalia ensiformis starch. *Journal of Agricultural and Food Chemistry* 45: 378–382.

Bhosale, R., & Singhal, R. 2006. Process optimization for the synthesis of octenyl succinyl derivative of waxy corn and amaranth starches. *Carbohydrate Polymers* 66: 521–527.

Bhosale, R., & Singhal, R. 2007. Effect of octenylsuccinylation on physicochemical and functional properties of waxy maize and amaranth starches. *Carbohydrate Polymers* 68: 447–456.

Blazek, J., & Copeland, L. 2009. Effect of monopalmitin on pasting properties ofwheat starches with varying amylose content. *Carbohydrate Polymers* 78: 131–136.

Cairns, P., Leloup, V., Miles, M. J., Ring, S. G., & Morris, V. J. 1990. Resistant starch: An X-ray diffraction study into the effect of enzymatic hydrolysis on amylose gels in vitro. *Journal of Cereal Science* 12: 203–206.

Caldwell, C. G., & Wurzburg, O. B. 1953. Polysaccharide derivatives of substituted dicarboxylic acids. *National Starch Products Inc.* USPTO, US 1953/2661349.

Carmona-Garcia, R., Sanchez-Rivera, M. M., Mendez-Montealvo, G., Garza-Montoya, B., & Bello-Perez, L. A. 2009. Effect of the cross-linked reagent type on some morphological, physicochemical and functional characteristics of banana starch (*Musa paradisiaca*). *Carbohydrate Polymers* 76: 117–122.

CFR (Code of Federal Regulation). 2001. Food starch modified. Food additives permitted in food for human consumption. Govt. Printing office, Washington, USA.21/1/172/172.892.

Chatakanonda, P., Varavinit, S., & Chinachoti, P. 2000. Relationship of gelatinization and recrystallization of cross-linked rice to glass transition temperature. *Cereal Chemistry* 77: 315–319.

Chen, X., He, X., & Huang, Q. 2014. Effects of hydrothermal pre treatment on subsequent octenyl succinic anhydride (OSA) modification of corn starch. *Carbohydrate Polymers* 101: 493–498.

Chi, H., Xu, K., Wu, X., Chen, Q., Xue, D., Song, C., Zhang, W., & Wang, P. 2007. Effect of acetylation on the properties of corn starch. *Food Chemistry* 106: 923–928.

Chung, H., Lee, S., Han, J., & Lim, S. 2010. Physical properties of dry-heated octenyl succinylated waxy corn starches and its application in fat-reduced muffin. *Journal of Cereal Science* 52: 496–501.

Chung, H. J., Woo, K. S., & Lim, S. T. 2004. Glass transition and enthalpy relaxation of cross-linked corn starches. *Carbohydrate Polymers* 55: 9–15.

Codex, F. C. 1996. *Committee on Food Chemicals Codex. Food and Nutrition Board, Institute of Medicine, National Academy of Sciences.* Published: National Academy Press, Washington, DC.

Colussi, R., Pinto, V. Z., El Halal, S. L. M., Vanier, N. L., Villanova, F. A., et al. 2014. Structural, morphological, and physicochemical properties of acetylated high, medium and low-amylose rice starches. *Carbohydrate Polymers*. 103: 405–413.

Fannon, J. E., Gray, J. A., Gunawan, N., Huber, K. C., & BeMiller, J. N. 2004. Heterogeneity of starch granules and the effect of granule channelization on starch modification. *Cellulose* 11: 247–254.

Fornal, J., Sadowska, J., Blaszczak, W., Jelinski, T., Stasiak, M., Molenda, M., & Hajnos, M. 2012. Influence of some chemical modifications on the characteristics of potato starch powders. *Journal of Food Engineering* 108: 515–522.

Garcia, F. P., Mendez, J. P., Marzo Mendez, M. A., Bello Perez, L. A., & Roman Gutierrez, A. D. 2012. Modification and chemical characterization of barley starch. *International Journal of Applied Science and Technology* 2: 30–39.

Garcia, V., Colonna, P., Lourdin, D., Buleon, A., Bizot, H., & Ollivon, M. 1996. Thermal transitions of cassava starch at intermediate water contents. *Journal of Thermal Analysis* 47: 1213 1228.

Gomes, A. M. M., Silva, C. E. M., & Ricardo, N. M. P. S. 2005. Effects of annealing on the physicochemical properties of fermented cassava starch (*polvilho azedo*). *Carbohydrate Polymers* 60: 1–6.

Gonzalez, Z., & Perez, E. 2002. Effect of acetylation on some properties of rice starch. *Starch/Stärke* 54: 148–154.

Gunaratne, A., & Hoover, R. 2002. Effect of heat–moisture treatment on the structure and physicochemical properties of tuber and root starches. *Carbohydrate Polymers* 49: 425–437.

Hadnadev, T. D., Hadnadev, M., Pojic, M., Rakita, S., & Krstonosic, V. 2015. Functionality of OSA starch stabilized emulsions as fat replacers in cookies. *Journal of Food Engineering* 167: 133–138.

Hirsch, J. B., & Kokini, J. E. 2002. Understanding the mechanism of cross-linking agents (POCl$_3$, STMP, and EPI) through swelling behaviour and pasting properties of cross-linked waxy maize starches. *Cereal Chemistry* 79: 102–107.

Hoover, R. 2000. Acid-treated starches. *Food Review International* 16: 369–392.

Hoover, R. 2010. The impact of heat–moisture treatment on molecular structures and properties of starches isolated from different botanical sources. *Critical Reviews in Food Science and Nutrition* 50: 835–847.

Hoover, R., & Hadziyev, D. 1981. Characterization of potato starch and its monoglyceride complexes. *Starch/Stärke* 33: 290–300.

Hoover, R., & Vasanthan, T. 1994a. The effect of annealing on the physicochemical properties of wheat, oat, potato and lentil starches. *Journal of Food Biochemistry* 17: 303–325.

Hoover, R., & Vasanthan, T. 1994b. Effect of heat–moisture treatment on the structure and physicochemical properties of cereal, legume, and tuber starches. *Carbohydrate Research* 252: 33–53.

Hoover, R., Swamidas, G., Kok, L. S., & Vasanthan, T. 1996. Composition and physicochemical properties of starch from pearl millet grains. *Food Chemistry* 56: 355–367.

Hormdok, R., & Noomhorm, A. 2007. Hydrothermal treatments of rice starch for improvement of rice noodle quality. *LWT-Food Science and Technology* 40: 1723–1731.

Huang, Q., Xiong Fu., Xiao-wei He, Fa-xing L., Yu, S., & Lin, Li. 2010. The effect of enzymatic pretreatments on subsequent octenyl succinic anhydride modifications of corn starch. *Food Hydrocolloids* 24: 60–65.

Hui, R., Chen, Qi-he., Ming-liang, F., Xu, Q., & Guo-qing, H. 2009. Preparation and properties of octenyl succinic anhydride modified potato starch. *Food Chemistry* 114: 81–86.

Jacobs, H., Eerlingen, R. C., Clauwaert, W., & Declour, J. A. 1995. Influence of annealing on the pasting properties of starches from varying botanical sources. *Cereal Chemistry* 72: 480–487.

Jayakody, L., & Hoover, R. 2008. Effect of annealing on the molecular structure and physicochemical properties of starches from different botanical origins-A review. *Carbohydrate Polymers* 74: 691–703.

JECFA 2001. Starches; Summary of evaluations performed by the Joint FAO/WHO expert committee on food additives.

Jenkins, P. J., & Donald, A. M. 1997. The effect of acid hydrolysis on native starch granule structure. *Starch/Starke* 49: 262–267.

Jeon, Y., Lowell, A. V., & Gross, R. A. 1999. Studies of starch esterification: reactions with alkenyl succinates in aqueous slurry systems. *Starch/Stärke* 51: 90–93.

Jiping, P., Shujun, W., Jinglin, Y., Hongyan, L., Jiugao, Y., & Wenyuan, G. 2007. Comparative studies on morphological and crystalline properties of B-type and C-type starches by acid hydrolysis. *Food Chemistry* 105: 989–995.

Jiranuntakul, W., Puttanlek, C., Rungsardthong, V., Puncha-arnon, S., & Uttapap, D. 2011. Microstructural and physicochemical properties of heat-moisture treated waxy and normal starches. *Journal of Food Engineering* 104: 246–258.

Juansang, J., Puttanlek, C., Rungsardthong, V., Puncha-Arnon, S., Jiranuntakul, W., & Uttapap, D. 2015. Pasting properties of heat-moisture treated canna starches using different plasticizers during treatment. *Carbohydrate Polymers* 122: 152–159.

Jyothi, A. N., Moorthy, S. N., & Rajasekharan, K. N. 2006. Effect of cross-linking with epichlorohydrin on the properties of cassava (*Manihot esculenta Crantz*) starch. *Starch/Stärke* 58: 292–299.

Kang, K. J., Kim, K., & Kim, S. K. 1994. Three-stage hydrolysis pattern of rice starch by acid treatment. *Journal of Applied Glycoscience* 41: 201–204.

Kaur, L., Singh, J., & Singh, N. 2006. Effect of cross-linking on some properties of potato (*Solanum tuberosum L.*) starches. *Journal of the Science of Food and Agriculture* 86: 1945–1954.

Kerr, R. W. 1952. Degradation of corn starch in the granule state by acid. *Stärke* 4: 39.

Kim, B., & Yoo B. 2010. Effects of cross-linking on the rheological and thermal properties of sweet potato starch. *Starch/Stärke* 62: 577–583.

Klein, B., Pinto, V. Z., Vanier, N. L., Zavareze, E. R., Colussi R, Do Evangelho, J. A., Gutkoski, L. C., & Dias, A. R. 2013. Effect of single and dual heat-moisture treatments on properties of rice, cassava, and pinhao starches. *Carbohydrate Polymers* 98: 1578–1584.

Koo, S. H., Lee, K. Y., & Lee, H. G. 2010. Effect of cross-linking on the physicochemical and physiological properties of corn starch. *Food Hydrocolloids* 24: 619–625.

Kurakake, M., Akiyama, Y., Hagiwara, H., & Komak, T. 2009. Effects of cross-linking and low molecular amylose on pasting characteristics of waxy corn starch. *Food Chemistry* 116: 66–70.

Lawal, M. V., Odeniyi, M. A., & Itiola, O. A. 2015. Material and rheological properties of native acetylated and pregelatinized forms of corn, cassava, and sweet potato starches. *Starch/Stärke* 67: 964–975.

Lawal, O. S. 2004. Succinyl and acetyl starch derivatives of a hybrid maize: Physicochemical characteristics and retrogradation properties monitored by differential scanning calorimetry. *Carbohydrate Research* 33: 2673–2682.

Lawal, O. S. 2012. Succinylated Dioscorea cayenensis starch: Effect of reaction parameters and characterisation. *Starch/Stärke* 64: 145–156.

Lawal, O. S., Adebowale, K. O., Ogunsanwo, B. M., Barba, L. L., & Ilo, N. S. 2005. Oxidized and acid thinned starch derivatives of hybrid maize: functional characteristics, wide-angle X-ray diffractometry and thermal properties. *International Journal of Biological Macromolecules* 35(1–2): 71–79.

Lawal, O. S., Lapasin, R., Bellich, B., Olayiwola, T. O., Cesaro, A., Yoshimura, M., et al. 2011. Rheology and functional properties of starches isolated from five improved rice varieties from West Africa. *Food Hydrocolloids* 25: 1785–1792.

Lee, C. J., Kim, Y., Choi, S. J., & Moon, T. W. 2012. Slowly digestible starch from heat-moisture treated waxy potato starch: Preparation, structural characteristics, and glucose response in mice. *Food Chemistry* 133: 1222–1229.

Lee, C. J., & Moon, T. W. 2015. Structural characteristics of slowly digestible starch and resistant starch isolated from heat–moisture treated waxy potato starch. *Carbohydrate Polymers* 125: 200–205.

Lim, S. T., Chang, E. H., & Chung, H. J. 2001. Thermal transition characteristics of heat– moisture treated corn and potato starches. *Carbohydrate Polymers* 46: 107–115.

Lim, S. T., & Seib P. A. 1993. Location of phosphate esters in a wheat starch phosphate by [31]P-nuclear magnetic resonance spectroscopy. *Cereal Chemistry* 70: 145–152.

Liu, H., Ramsden, L., & Corke, H. 1997. Physical properties and enzymatic digestibility of acetylated, ae, wx and normal maize starch. *Carbohydrate Polymers* 34: 283–289.

Liu, Z., Yin, li., Cui, F., Ping, L., Song, J., Ravee, V., Liqun, J., Yaping, X., Jianmiao, X., Geng, Li., Wang, Y., & Zheng, Y. 2008. Production of octenyl succinic anhydride-modified waxy corn starch and its characterization. *Journal of Agricultural and Food Chemistry* 56: 11499–11506.

Maache-Rezzoug, Z., Zarguili, I., Loisel, C., Queveau, D., & Buleon, A. 2008. Structural modifications and thermal transitions of standard maize starch after DIC hydrothermal treatment. *Carbohydrate Polymers* 74: 802–812.

Mason, W. R. 2009. Starch use in foods. In J. N. BeMiller & R. L. Whistler (eds), *Starch. Chemistry and Technology* 745–795. New York: Elsevier Inc.

Mbougueng, P. D., Tenin, D., Scher, J., & Tchiegang, C. 2012. Influence of acetylation on physicochemical, functional and thermal properties of potato and cassava starches. *Journal of Food Engineering* 108: 320–326.

Mirmoghtadaie, L., Kadivar, M., & Shahedi, M. 2009. Effects of cross linking and acetylation on oat starch properties. *Food Chemistry* 116: 709–713.

Moin, A., Ali, T. M., & Hasnain, A. 2016. Effect of succinylation on functional and morphological properties of starches from broken kernels of Pakistani Basmati and Irri rice cultivars. *Food Chemistry* 191: 52–58.

Morrison, W. R., Tester, R. F., Snape, C. E., Law, R., & Gidley, M. J. 1993. Swelling and Gelatinization of Cereal Starches. IV. Some Effects of Lipid-Complexed Amylose and Free Amylose in Waxy and Normal Barley Starches. *Cereal Chemistry* 70: 385–389.

Murphy, P. 2000. Starch. In G. O. Phillips & P. A. Williams (eds), *Handbook of Hydrocolloids* 41–65. Boca Raton, FL: CRC Press.

Olayinka, O. O., Adebowale, K. O., & Olu-Owolabi, B. I. 2008. Effect of heat moisture treatment on physicochemical properties of white sorghum starch. *Food Hydrocolloids* 22: 225–230.

Olayinka, O. O., Olu-Owolabi, B. I., & Adebowale, K. O. 2011. Effect of succinylation on the physicochemical, rheologial, thermal, and retrogradation properties of red and white sorghum starches. *Food Hydrocolloids* 25: 515–520.

Ortega-Ojeda, F. E., & Eliasson, A. C. 2001. Gelatinisation and retrogradation behaviour of some starch mixtures. *Starch/Stärke* 53: 520–529.

Park, S., Chung, M., & Yoo, B. 2004. Effect of octenyl succinylation on rheological properties of corn starch pastes. *Starch/Stärke* 56: 399–406.

Perez, E., Bahanassey, Y. A., & Breene, W. M. 1993. Some chemical, physical and functional properties of native and modified starches of amaranthus hypochondiracus and amaranthus cruentus. *Starch/Stärke* 45: 215–220.

Pinto, V. Z., Vanier, N. L., Klein, B., Zavareze, E. D. R., Elias, M. C., Gutkoski, L. C., et al. 2012. Physicochemical, crystallinity, pasting and thermal properties of heat-moisture-treated pinhao starch. *Starch/Stärke*. 64: 855–863.

Ratnayake, W. S., Hoover, R., & Warkentin, T. 2002. Pea starch: composition, structure and properties–A review. *Starch/Stärke* 54: 217–234.

Ratnayake, W. S., & Jackson, D. S. 2008. Phase transition of cross-linked and hydroxyl propylated corn (*Zea mays* L.) starches. *LWT-Food science and Technology* 41: 346–358.

Rodrıguez-Marın, M. L., Nunez-Santiago, C., Wang, Y., & Bello-Perez, L. A. 2010. Physicochemical and structural characteristics of cross-linked banana starch using three cross-linking reagents. *Starch/Stärke* 62: 530–537.

Saartrat, S., Puttanlek, C., Rungsardthong, V., & Uttapap, D. 2005. Paste and gelproperties of low-substituted acetylated canna starches. *Carbohydrate Polymers* 61: 211–221.

Sandhu, K. S., Sharma, L., & Kaur, M. 2015. Effect of granule size on physicochemical, morphological, thermal and pasting properties of native and 2-octenyl-1-ylsuccinylated potato starch prepared by dry heating under different pH conditions. *LWT-Food Science and Technology* 61: 224–230.

Sandhu, K. S., Singh, N., & Lim, S. T. 2007. A comparison of native and acid thinned normal and waxy corn starches: Physicochemical, thermal, morphological and pasting properties. *LWT-Food Science and Technology* 40: 1527–1536.

Segura-Campos, M., Chel-Guerrero, L., & Betancur-Ancona, D. 2008. Synthesis and partial characterization of octenylsuccinic starch from *Phaseolus lunatus*. *Food Hydrocolloids* 22: 1467–1474.

Sha, X. S., Xiang, Z. J., Bin, L., Jing, L., Bin, Z., Jiao, Y. J., et al. 2012. Preparation andphysical characteristics of resistant starch (type 4) in acetylated indica rice. *Food Chemistry* 134: 149–154.

Shaikh, M., Ali, T. M., & Hasnain, A. 2015. Post succinylation effects on morphological, functional and textural characteristics of acid-thinned pearl millet starches. *Journal of Cereal Science* 63: 57–63.

Sharma, M., Yadav, D. N., Singh, A. K., & Tomar, S. K. 2015. Rheological and functional properties of heat moisture treated pearl millet starch. *Journal of Food Science and Technology* 52: 6502–6510.

Sharma, M., Singh, A. K., Yadav, D. N., Arora, S., & Vishwakarma, R. K. (2016). Impact of octenyl succinylation on rheological, pasting, thermal and physicochemical properties of pearl millet (*Pennisetum typhoides*) starch. *LWT-Food Science and Technology* 73: 52–59.

Shi, Y. C., & Seib, P. A. 1992. The structure of four waxy starches related to gelatinization and retrogradation. *Carbohydrate Research* 227: 131–145.

Shogren, R. L., Vishwanathan, A., Felker, F., & Gross, R. A. 2000. Distribution of octenyl succinate groups in octenyl succinic anhydride modified waxy maize starch. *Starch/Stärke* 52: 196–204.

Shon, K., & Yoo, B. 2006. Effect of acetylation on rheological properties of rice starch. *Starch/Stärke* 58: 177–185.

Simsek, S., Ovando-Martínez, M., Whitney, K., & Bello-Perez, L. A. 2012. Effect of acetylation, oxidation and annealing on physicochemical properties of bean starch. *Food Chemistry* 134: 1796–1803.

Singh, A. V., & Nath, L. K. 2012. Synthesis and evaluation of physicochemical properties of cross-linked sago starch. *International Journal of Biological Macromolecules* 50: 14–18.

Singh, H., Chang, Y. H., Lin, J., Singh, N., & Singh, N. 2011. Influence of heat–moisture treatment and annealing on functional properties of sorghum starch. *Food Research International* 44: 2949–2954.

Singh, H., Sodhi, N. S., & Singh, N. 2009. Structure and functional properties of acid thinned sorghum starch. *International Journal of Food Properties* 12: 713–725.

Singh, H., Sodhi, N. S., & Singh, N. 2012. Structure and functional properties of acetylated sorghum starch. *International Journal of Food Properties* 15: 312–325.

Singh, J., Kaur, L., & McCarthy, O. J. 2007. Factors influencing the physico-chemical, morphological, thermal and rheological properties of some chemically modified starches for food applications – a review. *Food Hydrocolloids* 21: 1–22.

Singh, J., Kaur, L., & Singh, N. 2004. Effect of acetylation on some properties of corn and potato starches. *Starch/Stärke* 56: 586–601.

Singh, N., Chawla, D., & Singh, J. 2004. Influence of acetic anhydride on physicochemical, morphological and thermal properties of corn and potato starch. *Food Chemistry* 86: 601–608.

Siroha, A. K., & Sandhu, K. S. 2018. Physicochemical, rheological, morphological, and in vitro digestibility properties of cross-linked starch from pearl millet cultivars. *International Journal of Food Properties* 21: 1371–1385.

Siroha, A. K., Sandhu, K. S., Kaur, M., & Kaur, V. 2019. Physicochemical, rheological, morphological and in vitro digestibility properties of pearl millet starch modified at varying levels of acetylation. *International Journal of Biological Macromolecules* 13: 1077–1083.

Sodhi, N. S., & Singh, N. 2005. Characteristics of acetylated starches prepared using starches separated from different rice cultivars. *Journal of Food Engineering* 70: 117–127.

Song, X., Zhu, W., Li, Z., & Zhu, J. 2010. Characteristics and application of octenyl succinic anhydride modified waxy corn starch in sausage. *Starch/Stärke* 62: 629–636.

Sui, Z., Yao, T., Zhao, Y., Ye, X., Kong, X., & Ai, L. 2015. Effects of heat–moisture treatment reaction conditions on the physicochemical and structural properties of maize starch: Moisture and length of heating. *Food Chemistry* 173: 1125–1132.

Sun, S., Zhang, G., & Ma, C. 2016. Preparation, physicochemical characterization and application of acetylated lotus rhizome starches. *Carbohydrate Polymers* 135: 10–17.

Tester, R. F., & Morrison, W. R. 1990. Swelling and gelatinization of cereal starches. I. Effect of amylopectin, amylose and lipids. *Cereal Chemistry* 67: 551–557.

Tester, R. F., & Debon, S. J. (2000). Annealing of starch—a review. *International journal of biological macromolecules* 27(1): 1–12.

Thirathumthavorn, D., & Charoenrein, S. 2005. Thermal and pasting properties of acid-treated rice starches. *Starch/Stärke* 57: 217–222.

Thirathumthavorn, D., & Charoenrein, S. 2006. Thermal and pasting properties of native and acid-treated starches derivatized by 1-octeyl succinic anhydride. *Carbohydrate Polymers* 66: 258–265.

Van Hung, P., & Morita, N. 2005. Effect of granule sizes on physicochemical properties of cross-linked and acetylated wheat starches. *Starch/Stärke* 57: 413–420.

Van Hung, P., Vien, N. L., & Phi Nguyen, T. L. 2016. Resistant starch improvement of rice starches under a combination of acid and heat-moisture treatments. *Food Chemistry* 191: 67–73.

Vasanthan, T., Sosulski, F. W., & Hoover, R. 1995. The reactivity of native and autoclaved starches from different origins towards acetylation and cationization. *Starch/Stärke* 4: 135–143.

Waduge, R. N., Hoover, R., Vasanthan, T., Gao, J., & Li, J. 2006. Effect of annealing on the structure and physicochemical properties of barley starches of varying amylose content. *Food Research International* 39: 59–77.

Wang, L., & Wang, Y. J. 2001. Structures and physicochemical properties of acid-thinned corn, potato and rice starches. *Starch/Stärke* 53: 570–576.

Wang, S., & Copeland, L. 2015. Effect of acid hydrolysis on starch structure and functionality: A review. *Critical Reviews in Food Science and Nutrition* 55: 1081–1097.

Shubeena, S., Wani, I. A., Gani, A., Sharma, P., Wani, T. A., Masoodi, F. A., Hamdani, A., & Muzafar, S. 2015. Effect of acetylation on the physico-chemical properties of Indian horse chestnut (*Aesculus indica* L.) starch. *Starch/Stärke* 6: 311–318.

Wani, I. A., Sogi, D. S., & Gill, B. S. 2012. Physicochemical properties of acetylated starches from some Indian kidney bean (*Phaseolus vulgaris L.*) cultivars. *International Journal of Food Science and Technology* 47: 1993–1999.

Watcharatewinkul, Y., Puttanlek, C., Rungsardthong, V., & Uttapap, D. 2009. Pasting properties of a heat-moisture treated canna starch in relation to its structural characteristics. *Carbohydrate Polymers* 75: 505–511.

Wurtzburg, O. B. 1960. Preparation of starch derivatives. US Patent No. 2,935,510.

Wurzburg, O. B. 1978. Starch, modified starch and dextrin. In *Products of the Corn Refining Industry; Seminar proceedings* 22–32. Washington, DC: Corn Refiners Association, Inc.

Yoneya, T., Ishibashi, K., Hironaka, K., & Yamamoto, K. 2003. Influence of cross-linked potato starch treated with $POCl_3$ on DSC, rheological properties and granule size. *Carbohydrate Polymers* 53: 447–457.

Zavareze, E. R., & Dias, A. R. G. 2011. Impact of heat–moisture treatment and annealing in starches: A review. *Carbohydrate Polymers* 83: 317–328.

Zhang, J., Wang, Z., & Yang, J. 2010. Physicochemical properties of canna edulis ker starch on heat-moisture treatment. *International Journal of Food Properties* 13: 1266–1279.

Zheng, G. H., Han, H. L., & Bhatty, R. S. 1999. Functional properties of cross-linked and hydroxypropylated waxy hull-less barley starches. *Cereal Chemistry* 76: 182–188.

# 7 Biotechnological Applications for Improvement of the Pearl Millet Crop

*Supriya Ambawat, Subaran Singh, Shobhit, R.C. Meena and C. Tara Satyavathi*

## CONTENTS

## 7.1 INTRODUCTION

Pearl millet [*Pennisetum glaucum* (L.) R. Br.] is a hardy cereal and a staple food cultivated on more than 26 mha in arid and semi-arid tropical regions of Africa, Asia and Latin America. This is generally cultivated in harsh climatic environments where other crops, such as sorghum and maize, are not able to generate good yields. It is an annual, small-seeded, highly cross-pollinated, diploid ($2n = 2x = 14$), $C_4$ plant having high photosynthetic efficiency. It has least demand in terms of input and renders excellent nutritional values and biomass production potential. In India, it is the fourth most extensively cultivated crop, with an average production of 9.73 mt

**115**

(Directorate of Millet Development, 2018). It has multiple uses in the form of food, feed, fodder, forage, biofuel, brewing and building material. It is rightly termed as a nutri-cereal, as it has high nutrition value and is rich in carbohydrate, dietary fiber, protein, essential fatty acids, vitamins and minerals such as calcium, zinc, iron, potassium and magnesium. It can have several health benefits, such as reducing blood pressure, and alleviating thyroid, diabetes, cardiovascular and celiac diseases. However, its direct consumption as food has extensively declined during the past three decades due to various reasons. In this situation, it is important to disseminate to people its nutritional value, and efforts should be accelerated to generate its demand through quality improvement and value-addition, which is possible by exploring the pearl millet's genetic features using advanced technologies.

More than 500 million nutritionally undernourished and poorest people of the world obtain food and nutritional security by using pearl millet. During the past 20 years, pearl millet production has increased due to acceptance of hybrids in India as well as an increase in its production area in West Africa, but different abiotic and biotic stresses produce a big obstacle to its use and lead to a large reduction in yields. Germplasm conservation and pearl millet genetic improvement depends on knowledge of genomic diversity and composition of population. Thus, characterization of genetic diversity within diverse collections of pearl millet is required for their proficient utilization (Varshney et al., 2009). Pearl millet has unique characteristics which differentiates it from other cereals. It can be grown under dry conditions in poor soils with fewer inputs. Pearl millet has high polymorphism due to traditional cultivation in stressful environmental conditions, and is highly resistant to adverse nature and independent domestication events. This high polymorphism has been exposed by genome mapping, which has further helped in tagging and mapping of different genomic loci that offer several useful traits. Various strategies were employed for pearl millet's genetic improvement, thus increasing crop productivity, which will ultimately provide food and nutritional security in a changing climatic scenario. In addition to conventional approaches, different modern techniques, such as genetic modifications or transgenics, can also help in the genetic improvement of millet. In the past, genomic research into pearl millet has lagged behind due to several reasons. However, numerous research efforts have now been made in genomics, yielding much information and the generation of a huge amount of genomic resources such as recently revealed genome sequence and re-sequencing of approximately 1,000 lines, which defines the entire diversity (Varshney et al., 2017).

Targeted novel approaches, such as marker–trait associations, tools and techniques of genomic selection, genome sequences, NGS and GBS, can accelerate genetic gains but they have been used in a limited way in orphan crops. Thus, improvement and application of several high throughput genomic tools in pearl millet is required to advance the breeding efficiency of conventional approaches and this is the best way to reduce the effect of the changing climate on its production. Presently available genetic and genomic resources and recent biotechnological advances can play a vital role in accelerating the rate of genetic gains and, thus, provide good scope for future crop improvement programs. A complex quantitative trait such as drought tolerance has been studied extensively with the application of molecular tools. Advanced techniques such as NGS and molecular profiling have made it easy to identify and tag the gene of interest, as they are efficient in tracing DNA sequence variations in thousands of genomic regions and nowadays they are readily affordable (Elshire et al., 2011). The extent of polymorphism present in pearl millet was detected through genome mapping, which will ultimately encourage mapping and gene tagging of various significant traits, including grain, fodder yield, drought tolerance, etc. In addition, gene introgression into useful genetic backgrounds, using an advanced tool such as MAS, has helped greatly in crop improvement by reducing prolonged phenotypic estimation and selection (Vadez, Hash and Kholova, 2012).

Molecular tools and genomic studies are gaining lot of momentum these days and playing a major role in crop improvement programs, as they have many uses in improving the efficiency and accuracy of conventional breeding. Integrated knowledge of genomics, as well as transcriptomics, proteomics and metabolomics, may be beneficial for advancements in the biofortification of pearl millet. This chapter reviews the importance and interventions of various biotechnological approaches, such as transgenic approaches, molecular tools and genomic studies into nutri-cereal pearl millet to improve production and productivity as well as quality.

## 7.2 CONVENTIONAL APPROACHES FOR GENETIC IMPROVEMENT

During the last several decades, there has been much advancement in pearl millet genetic improvement in India. Earlier, increased production was achieved through the use of controlled hybrids and conventional breeding methods of selection. A genetic improvement program in India was started in the 1930s by selecting local landraces and then developing high-yielding hybrids that were resistant to diseases and tolerant of abiotic stresses such as heat and drought. Initially, concentration was focused on increasing the yield by progeny testing and mass selection, which resulted in the development of several open-pollinated varieties (OPVs). But it resulted in only partial improvement in yield, due to limited field testing and, moreover, the OPVs were generated from landraces having a narrow genetic base. Hybrid development in India has been accomplished in three conspicuous phases, as described by Yadav and Rai (2013). Several genetically diverse hybrids were developed using diverse combinations of phenotypic traits related to different ecological regions with germplasm from the Indian subcontinent and Africa. During the past 25 years, more than 150 cultivars have been released, having different combinations of diverse phenotypic traits, providing many options to farmers from different production ecologies.

Several recurrent selection methods (mass selection, S1 and S2 progeny selection, restricted recurrent phenotypic selection, gridded-mass selection, half-sib selection, full-sib selection) have been used in different population improvement programs that had varying achievements in genetic improvement for different composites (Singh et al., 1988; Zaveri, Phul and Dhillon, 1989). Diverse trait-based composites, such as medium composite, dwarf composite, late composite, early composite, high tillering composite, etc., having a wide genome base, were developed until the late 1980s (Rai and Kumar, 1994). Pearl millet, being a highly cross-pollinated crop, possesses a higher degree of heterosis for stover, grain and yield and so efforts were made in the 1950s to utilize heterosis. However, due to lower yield advantages in comparison to OPVs, a narrow adaptation range and smaller number of seed production programs, they were not much admired. Hybrids can be developed by selecting two parental lines having the desired traits and crossing those using conventional breeding methods. Genetic variation among target traits within the gene pool is a decisive factor in the success of a crop improvement program. It further depends on environmental conditions and is, thus, determined by GXE interactions. Therefore, GXE interactions are crucial aspects for the establishment of high-nutrition and stable cultivars and must be taken into account (Moghaddam and Pourdad, 2009). Multi-environmental trials can, therefore, be helpful for confirming the strength of phenotypic data. The existence of GEI can decrease the strength of any kind of analysis, limit the importance of results and restrict the competence of the elite genotype selection (Gurmu, Mohammed and Alemaw, 2009).

Pearl millet may be a good genomic resource to isolate different biotic/abiotic stress tolerant candidate genes, thus leading to its genetic improvement along with some other crops (Vadez, Hash and Kholova, 2012). On the other hand, pearl millet has good grain qualities, including a higher amount of essential amino acids, vitamins and minerals, but their bioavailability must be further improved by using novel promoters or reducing antinutrients. Conventional approaches are time consuming and need the help of modern biotechnological tools to accelerate millet development programs. Recently reported genome sequence information can appreciably accelerate gene innovation and trait mapping, thus improving the perception of several complicated gene pathways and their interactions. An enormous development in the field of genomics during past years has rendered available various novel tools for precise and faster breeding programs. Huge marker data sets are used for genomics-assisted breeding instead of one or a few loci linked with the trait and, thus, it will be proved to be more useful in the coming years for the improvement of a resource poor crop like pearl millet.

## 7.3  BIOTECHNOLOGICAL APPROACHES

Although conventional plant breeding, along with good crop management practices, has removed several agronomic constraints which reduce crop yield and effect nutritional quality, there may be several aspects where the full potential of existing genetic resources has not been utilized fully. Therefore, advanced plant breeding strategies need the support of genomics and genetic transformation technologies which can utilize successful gene-based techniques for improving abiotic and biotic stress tolerance in millet and can contribute to an advancement of sustainable agriculture for dry regions. A conventional approach is good, but is time consuming and, for enhanced production and improvement, we need to characterize pearl millet genetically using several biotechnological tools and techniques. In addition to bioinformatics and systems biology, diverse "omics" approaches such as transcriptomics, proteomics and metabolomics can be used for quantitative and qualitative analysis of gene expression that will allow more specific use of MAS and transgenic technologies (Dita et al., 2006).

Molecular approaches can be utilized to portray and control crop genomes in order to better identify different genes and biochemical pathways governing several characteristics such as yield, abiotic stress tolerance, disease resistance, insect–pest resistance, plant architecture and quality of produce. In this context, huge investments have been made in the area of biotechnology for the major crops of wheat, rice and maize during past decades, but fewer pains have been taken in regard to pearl millet and still the efforts in this direction for this crop are in the initial stages. DNA-based approaches can be useful primarily in extensive areas of molecular breeding: viz., marker-assisted breeding which aims to increase the power of genetic analysis by using markers and transgenics, which involves gene transfer from a genotype to another genotype, generally known as "genetic engineering," and eventually leads to the addition of novel traits to the crop. Molecular markers act as a major tool in molecular breeding for the identification of useful variants, genetic diversity conservation and precise selection of genotypes for further use in conventional plant breeding (Bollam et al., 2018)

Some molecular techniques can be useful to any species, despite the lack of availability of a DNA sequence, and can be performed in any laboratory having even a minimum level of supplies and equipment. The aim behind it is to effectively utilize natural resources, develop biotic/abiotic stress resistance/tolerance and improve quality in order to facilitate wider access. Genetic characterization studies are lagging behind in respect of pearl millet in comparison to other cereals and its improvement using biotechnology was ignored mainly due to economic or regional reasons. Genetic maps, NGS, GBS, GWAS, synteny studies, expression profiling, fine QTL mapping, candidate gene identification and genetic engineering technology are some of the useful platforms which can be used for the advancement of nutrient rich pearl millet. A few efforts have been made in order to develop molecular tools for pearl millet. Several initial studies were performed to comprehend the genetic diversity of pearl millet, using molecular markers such as RFLP, AFLP and SSR (Allouis et al., 2001; Liu et al., 1994; Qi et al., 2004). Molecular markers, sequence information and genetic maps and are some of the basic genomic resources required for genetic exploration and molecular breeding approaches.

Genomics is a useful branch of biotechnology that deals with the development of molecular markers, identification of putative QTLs, uses advanced genomic tools and MAS for the development and improvement of useful cultivars. These genomic tools help to smooth the progress of breeding strategies for crop improvement programs by reducing the labor intensive and time consuming direct screening of germplasm grown under field, as well as greenhouse, conditions. A number of major QTLs effecting abiotic stress tolerance in millet were identified and mapped as a major QTL for terminal drought tolerance. Thus, genome-wide understanding of millet germplasm is required to obtain full information and explore the different genomic resources in this crop. Further, advanced biotechnological approaches such as omics could serve as the most available potential strategies for improving the pearl millet crop. Omics includes several biotechnological applications such as genomics, functional genomics, gene expression, transcriptomics, proteomics and metabolomics, which leads to significant alterations in the plant transcriptome, proteome and metabolome (Ahuja et al., 2010). NGS technologies have made it easy to sequence many crops and use them in genetic map development. The sequencing method can be attached and used to genotype large populations by using GBS and restriction site-associated DNA sequencing (RADseq) approaches which are based on the reduced representation sequencing technique (Elshire et al., 2011; Poland et al., 2012). RNAi silencing and genome editing tools are some of the recently developed biotechnological techniques which can play a key role in pearl millet improvement, and methods for implementing these are still being developed in pearl millet.

### 7.3.1 TISSUE CULTURE AND TRANSGENIC/TRANSFORMATION APPROACHES

Regeneration using tissue culture technique and transgenics are other strategies which can be significantly used for genetic improvement in pearl millet. They help in obtaining multiple identical copies of disease- and pest-free plants, along with regeneration of the entire plant from transformed tissues. The efficiency of regeneration mainly decides the success rate in plant transformation. Thus, an optimum regeneration method should be fully established for each ecotype and plant species in order to achieve plant transformation. Different morphogenic pathways have been studied for tissue culture and plant regeneration in millet. Gene transfer experiments have also been carried out using several methods, but transgenic plants were established in Bahiagrass and pearl millet only due to lower economic importance and, moreover, the major focus of research is on the improvement of major cereals such as wheat, maize and rice (Girgi et al., 2002; Smith et al., 2002).

Several reviews are available on plant regeneration in millet (Kothari and Chandra, 1995; Repellin et al., 2001). Efforts are being made to include a tertiary gene pool into improved cultivars via genetic transformation protocols for millets. However, millet is not very responsive to transformation protocols due to a lack of model genotypes that may be proficiently transformed for any millet species. Further, parameters affecting millet regeneration include explants, plant growth regulators (PGRs) and environmental factors such as temperature, pH and light. Although nutritional improvement of pearl millet grains by means of genetic engineering is an essential area of research to be focused on for nutritional security, still less effort was made in this direction. Different explants, such as whole seeds, seedling leaf bases, mature embryos, immature inflorescence and roots, have been used for regeneration experiments in millets for initiating cultures, but immature embryos with scutellum at the milk stage has been found to be the best starting material. Initially, callus is formed and, later, regeneration occurs via somatic embryogenesis or organogenesis. Historically, protoplasts were found to be useful for transformation because the desired DNA was introduced by means of chemical methods or electroporation, but protoplasts were not suitable for regeneration due to recalcitrance (Potrykus, 1990).

The introduction of DNA into intact cells and tissues using a particle gun was the preferred method of gene transfer. To develop transgenic plants, immature embryos were transformed with DNA coated particles. Gold/tungsten particles coated with the desired gene were introduced into cells via particle delivery devices. Several vectors were tested in several combinations of *Adh1*, CaMV 35s, *Emu* intron (maize *Adh1*) and terminators. Regeneration of hygromycin resistant pearl millet plants were noted by Lambé, Dinani and Deltour (2000) from type II callus.

## 7.3.2 RECOMBINANT DNA TECHNOLOGY

rDNA technology is a useful technique which plays a vital role in the development of pearl millet germplasm (O'Kennedy, Grootboom and Shewry, 2006) but this technique was mostly implemented in major cereals rather than pearl millet. Though current advances for the development of pearl millet were accomplished using conventional breeding methods and marker assisted selection, transgenics and *in vitro* culture may also help to expand the gene pool further. An embryogenic *in vitro* culture system was developed for the first time in pearl millet in the 1980s (Vasil and Vasil, 1981). Later, other reliable protocols for transformation and *in vitro* culture were generated in pearl millet that can be further improved upon. Direct gene transfer is another aspect that uses rDNA technology to insert foreign genes into the genome of other plants. Pearl millet transformation was first of all performed using particle bombardment, where transformation was carried out using plasmid pMON8678 (Taylor, Vasil and Vasil, 1991). Later, different groups used this approach in pearl millet using different target tissues, and transgenics were raised against fungal pathogens (Ceasar and Ignacimuthu 2009). But, so far, there is not any such report on developing abiotic stress tolerant pearl millet.

In another report, *Agrobacterium* mediated transformation was used in pearl millet shoot apices (Ignacimuthu and Kannan, 2013). This technique could be further implemented to pearl millet in order to increase the level of Fe accumulation and establish antinutrient compounds such as phytate. Native genes from closely related species can be used to alter a plant's metabolism by "genomics-guided transgenes" (GGT) via direct gene transfer (Strauss, 2003).

### 7.3.3 MOLECULAR MARKERS

A considerable advancement has been made during the past several years in the expansion of molecular tools for pearl millet in comparison to other orphan crops, and this will be helpful in enhancing the breeding of improved cultivars (Vadez, Hash and Kholova, 2012). DNA marker-based linkage maps, molecular markers, BACs and EST libraries are the landmark areas where significant progress has been observed. For pearl millet, several molecular markers, such as restriction fragment length polymorphisms (RFLPs) (Liu et al., 1994), sequence tagged sites (STSs) (Devos et al., 1995), amplified fragment length polymorphisms (AFLPs) (Allouis et al., 2001), single-stranded conformational polymorphism (SSCP) (Bertin, Zhu and Gale, 2005), gSSRs (simple sequence repeats) (Qi et al., 2004), EST-SSRs (expressed sequence tag–simple sequence repeats) (Senthilvel et al., 2008), diversity arrays technology (DArT) (Supriya et al., 2011), single nucleotide polymorphism (SNP), conserved intron spanning primer (CISP) markers (Sehgal et al., 2012 and SCAR marker (Jogaiah et al., 2014) have been developed in the past 25 years. Among these, SSR was found to be a powerful tool for marker assisted breeding (MAB) due to its abundance, reproducibility, co-dominance and variability. But, the comparatively high cost per data point and low coverage of the genome constrain its relevance in QTL mapping and marker assisted selection programs.

Initiation of the development of molecular markers for pearl millet improvement was set up in 1991 with the DFID-JIC-ICRISAT project (Gale et al., 2005). As a result, a linkage map was first constructed in pearl millet by means of the use of RFLPs (Liu et al., 1994). Earlier, genetic maps were very much based on morphological markers, but now, due to current advances in biotechnology, molecular markers are being used for the construction of high-density maps (Hyten and Lee, 2016). Nowadays, high throughput and cost-effective markers such as SNP and DArT are the markers of choice in molecular breeding used to construct high-density maps. DArT, a sequence-independent technique, is a good substitute for overcoming the limitations of gel-based marker systems. It is established by using a metagenome and provides high multiplexing for the diversity analysis of several genotypes (Yang et al., 2011). Its high throughput technique as a platform can be utilized for discovering, as well as scoring, polymorphism. It has been used for diversity assessment, genotyping and genetic mapping in many cereals and millets (Jing et al., 2009; Mace et al., 2008; Supriya et al., 2011). DArT-based linkage maps were first developed in pearl millet by Senthilvel et al. (2010) and Supriya et al. (2011). Development of NGS technology has resulted in a quick and inexpensive development of a high throughput marker system that enabled the genotyping of large mapping populations and the creation of ultra high-density maps. High-density genetic maps and molecular markers also help in comparative mapping and synteny studies in crop plants; for example, one of the recent studies has shown the synteny between foxtail millet and pearl millet chromosomes, along with those of other grasses (Rajaram et al., 2013). Recently, a modified GBS platform was used for development of a higher density genetic map, along with uniform distribution of markers, than previously available maps. Various efforts resulted in the generation of 3,321 SNPs (Moumouni et al., 2015) that will ultimately improve extensively the advancement of genes and QTL mapping in bi-parental populations and will provide panels for association analysis. The genetic maps have been demonstrated to be valuable not only for detection and breeding of promising QTLs for various traits, including grain and stover yield, components for drought adaptation and terminal drought tolerance, etc., but also for improved perception of multifarious associations between pearl millet and other cereals (Bidinger et al., 2007; Jones et al. 2002; Kholová et al., 2012; Yadav et al., 2003, 2004). Thus, advances in molecular markers can contribute to gene identification, tagging and detection of QTL for several uses in agriculture and fine QTL mapping is possible using high density markers which will also be helpful for molecular breeding applications and MAS (Wu et al., 2014).

The accessibility of extensive genome-wide markers in pearl millet can also help in the tagging of genes and QTLs related to nutrition, thus enhancing grain quality. Further, it is anticipated that, with increased understanding of genetic variations for micronutrients in the genome, it may be possible to discover and validate high micronutrient uptake genes that will ultimately result in improved varieties that can be used by the wider population and smallholders (Muthamilarasan et al., 2016). Discovery of SNPs and insertions and deletions will also help to identify candidate genes defining nutritional traits. Some of the newly established ESTs and genomic SSRs in pearl millet were proved to be useful in defining heterotic gene pools (Ramya, Ahamed and Srivastava, 2017). These outcomes, along with high-throughput genotyping and phenotyping for several biotic/abiotic stresses, may help in creating rapid development of improved hybrids and populations for different agro ecologies. With the identification of several high throughput SNPs, diversity analysis and population structure, genetic mapping, genome-wide association studies (GWAS) and identification of candidate genes controlling major drought-tolerant QTL became easier (Metzker, 2010; Sehgal et al., 2012). In other studies, thousands of SNPs were found in pearl millet by using the GBS technique (Hu et al., 2015; Moumouni et al., 2015). Recently, the inbred germplasm association panel (PMiGAP) of pearl millet, comprising 250 inbred lines for drought tolerance, was also used for GWAS, including three indel markers and 22 SNPs (Sehgal et al., 2015). Fe works have been performed in millet using markers and the genotyping platform to facilitate quality improvement in millet. In addition, gene coding for kinases, oxidases, transferases and different types of proteins, which are typically related to grain yield QTL, leaf rolling and flowering time under drought conditions, were also co-mapped with QTL for drought tolerance. Thus, it indicates the significance of functional markers in the identification of candidate genes and functional evaluation of indistinctly related genomes.

### 7.3.4 QTL Mapping

Several QTL mapping reports were observed in millet: grain yield, height, stover yield, biotic stress, such as downy mildew resistance, and abiotic stresses (Ambawat et al., 2016; Jones et al., 2002; Kannan et al., 2014; Yadav, Sehgal and Vadez, 2011) and some of the important markers, mapped traits and QTLs have been described by Bollam et al. (2018). Numerous QTLs were detected for downy mildew on different linkage groups against different pathotypes (Hash and Witcombe, 2001; Jones et al., 2002). Mapping of QTL for terminal drought tolerance produced a very significant finding in pearl millet to establish the physiological and genetic basis and provided a better way to improve drought tolerance and yield in water-limited environments (Yadav et al., 2004). In another study, approximately 60 putative downy mildew (DM) resistance QTLs have been found to be linked to specific DNA markers in millet (Breese et al., 2002). Many of them are useful for different Indian pathotypes of *S. graminicola* and were transferred to commercial R-lines and B-lines. Biofortified pearl millet varieties can also be developed by targeting candidate genes and high Fe and Zn QTLs. Further, the genomic positions of grain iron- and zinc-linked SSR markers are being identified in the consensus map and research is in progress towards mapping QTLs for flour rancidity.

### 7.3.5 Marker Assisted-Selection (MAS)

MAS is a tool used in molecular breeding to identify DNA sequences located near genes that may be used for breeding different traits and is much simpler for genes associated with large phenotypic effects such as disease resistance. It has been proved to be successful for germplasm enhancement and several other field applications for major crops. With advancing marker techniques, more applications of MAS may be anticipated, ultimately giving the prospective benefits of rapid release of useful varieties. The downy mildew-resistant variety, HHB 67, is the first successful popular hybrid of pearl millet cultivated in North India and has been developed by marker assisted breeding (MAB) (Hash et al., 2006). Several studies have proved that MAS has many positive cost benefits and, thus, is a very useful and effective technique for crop improvement (Bohn et al., 2001; Morris et al., 2003).

### 7.3.6 Omics Approaches

Genetic improvement of pearl millet has been increased at a higher rate by using existing genetic and bioinformatic resources, along with precision phenotyping. In this context, omics approaches are one of the most straightforward and biotechnological applications with the potential to improve pearl millet. With the ongoing genomics programs in pearl millet and the availability of genome sequencing, the relevance of omics in millet is very important. The current and upcoming high-throughput sequencing platforms can provide a wide variety of applications to researchers, such as SNPs, molecular markers and identification of small RNAs. Omics approaches can be widely used to analyze the genes responsible for stress adaptation and identification of various QTLs governing the adaptive response under stressful conditions. Further, it can help to predict the molecular mechanism of stress response/tolerance, disease resistance and nutritional quality by studying the gene, protein or metabolite profile and their phenotypic effects. NGS approaches, along with genome-wide expression profiling studies, can resolve the issues raised by large genomes, particularly those of pearl millet. In comparison to genomics and transcriptomics, proteomics and metabolomics studies are still lagging behind in pearl millet (Lata, 2015). But rapid advancement in high-throughput proteomics and metabolomics approaches, such as flow injection time-of-flight mass spectrometry etc., along with various omics strategies, will reform the study of complex biological systems of millet. Thus, overall, it will improve our basic understanding of the molecular basis of this crop by utilizing large-scale identification of genes/proteins and metabolites involved in multiple molecular and signaling networks. A concerted effort including all omics will be an important step towards understanding the different mechanisms in millet that can be utilized further for MAS or conventional breeding.

### 7.3.6.1  Genomic Tools

Several complex mechanisms like plant stress response, adaptation, survival and consequent yield are governed by different physiological factors and cellular molecular networks (Ahuja et al., 2010). Thus, understanding the molecular mechanism of stress response is essential to address and surmount the issue of changing climate, which ultimately leads to stability in yield. Conventional breeding approaches are very difficult for drought tolerance, as it depends on location-specific stress factors. Thus, the integration of advanced genomic technologies with breeding might be an excellent and effective strategy to explore and control different stress responses in pearl millet (Langridge and Fleury, 2011). Engineering for drought and heat tolerance in pearl millet, as well as other cereals, can be possible through identification of different gene families linked with these traits. Thus, genomics provides breeders with several new tools and techniques (Figure 7.1) to study the whole genome of plants and has become a necessary factor for breeding that will be valuable in identifying and understanding novel genes for different traits of stress tolerance and quality improvement. By studying the genotype and its relationship with the phenotype, more efficient cultivars can be developed, having the greatest potential to increase yield. MAS and introgression of a desired gene into a specific genetic background can assist crop improvement, as it reduces prolonged and cumbersome phenotypic screening and selection. Genomic methods can lead to tagging and genetic mapping of significant traits in the pearl millet genome and, with the progress in genotyping, different molecular plant breeding approaches have been developed for different crops. As millet is a climate-resilient crop, it can be an effective source of novel stress tolerant genes. Millet has close phylogenetic relationships with other cereals, thus enabling incorporation into other cereals of detected QTLs and novel alleles/ genes identified from millets for different characteristics to enhance food safety. Bollam et al. (2018) reviewed some useful genomic tools and markers linked with diverse traits in pearl millet. Genetics and functional genomics tools have helped to enhance productivity, nutritional quality and stability of different food crops over many years in a very effective way. Using genomic tools, now we can identify candidate genes and QTLs related to beneficial traits, and availability of diverse genetic resources and germplasm will assist this.

The gene banks of pearl millet contain varieties with high iron and zinc levels which may be helpful in developing new high-yielding and biofortified pearl millet varieties using MAS. A highly significant and positive correlation has been found between Zn and Fe, offering good chances of escalating the levels of both micronutrients and good yield (Velu et al., 2007). Thus, genomics approaches are very helpful to hasten the pace of development of biofortified varieties in millet in a very cost efficient way. Remarkable progress was achieved in pearl millet using genomic techniques, and many regulatory gene networks were also identified to understand associations with other cereals and between varieties of pearl millet (Devos and Gale, 2000). It has also been used to identify several QTLs responsible for drought adaptation, terminal drought tolerance, grain and stover yields (Kholová et al., 2012; Yadav et al., 2003, 2004). Numerous genetic maps were constructed for different traits and many promising QTLs were identified and incorporated in breeding programs for genetic gains (Bidinger et al., 2007; Jones et al., 2002; Serraj et al., 2005; Yadav, Sehgal and Vadez, 2011).

Recently, entire genome sequencing and re-sequencing of around 993 pearl millet lines that represent overall diversity were reported by an international consortium led by ICRISAT (Varshney et al., 2017). This is like a genetic blueprint that can be used easily to develop genomic tools and thus, in turn,

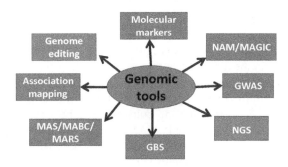

**FIGURE 7.1**  Various genomic approaches used for crop improvement

will accelerate the development of improved cultivars. During this research, nearly 30 million SNPs were reported in PMiGAP, where around 450,000 SNPs of high-quality were identified. Approximately 88,256 SSR markers were recognized from sequenced data and 74,891 SSR markers were used to design primers. These markers can be utilized in future genetics and breeding programs to comprehend trait variation effectively and speed up millet genetic improvement (Varshney et al., 2017). This will also be helpful for marker linked studies and candidate gene detection for yield and stress tolerance. Further, genome sequencing could be combined with transcriptomic analysis to establish genomic loci, gene functions and their expression patterns for different conditions and traits. Sequencing studies will be helpful in overcoming various biotic/abiotic stresses and increasing productivity. Thus, draft genome and re-sequencing data will provide great assistance and opportunities to the researchers in understanding the trait variation in an effective way, and will be useful for advancing genetic improvement of the crop. In another genomic study on pearl millet, a set of chromosome segment substitution lines have been developed for all seven linkage groups, which is very useful for analyzing the genetic effects of complex characteristics such as grain yield (Kumari et al. 2014).

### 7.3.6.2  Genome Wide Association Studies (GWAS)

GBS and phenotypic data are combined in order to identify genomic loci governing specific traits by using GWAS. It has emerged as a potent tool to explore the genetic makeup of many species, as it uses natural diversity accumulated by multi-generational recombination actions in the germplasm panel or population (Deschamps, Llaca and May, 2012). It increases mapping resolution in comparison to linkage mapping populations, thus accelerating crop improvement. GWAS acts as a catalyst during mining of candidate genes, which can be further confirmed through genetic transformation or T-DNA mutants, thus facilitating the genetic alterations or MAS of validated genes, finally resulting in improved varieties that are rich in nutrients (Huang and Han, 2014). A comparative study of millets and non-millets is possible through synteny investigation in order to identify common genes linked to nutrition biosynthesis. Once such genes, alleles and QTLs are identified, these could be integrated into elite lines using MAS or transgene methods. In this context, genomics, transcriptomics, proteomics, metabolomics, ionomics and bioinformatics are some recently developed approaches that can play a major role (Muthamilarasan et al., 2016). The genomic resources that characterized major cereals can also be helpful for pearl millet, because considerable genomic co-linearity exists in many cereal crops and, thus, will help to transfer genes from major crops to minor (Varshney, Hoisington and Tyagi, 2006). This, in turn, will be helpful in improving cereal biodiversity analysis and detection of valuable variants, MAS of alleles and allele combinations. Further, cloning and transfer of useful alleles may be carried out successfully among the cereal family. Assets derived from already sequenced crops will help to identify effectively several key genes, their pathways and specific functions (Chen et al., 2016). Thus, wide pearl millet germplasm is required to attain more precise and full potential of genomic resources effectively.

### 7.3.6.3 Functional Genomics

In the post genomics era, broad analysis using functional genomics tools such as gene silencing, insertional mutagenesis, overexpression and targeted induced local lesion in genome (TILLING) have played a major part in enhancing our knowledge of complex regulatory networks involved in stress response, adaptation and tolerance in plants. Different omics technologies are also generating massive data sets which can be exploited for identification of important candidate genes to be used in millet improvement programs. With the progression of genome sequencing projects and proteomics, functional validation of hundreds and thousands of identified genes or proteins becomes a big challenge which can be resolved using reverse genetics strategies. Thus, overexpression of a gene in the same or heterologous system under the control of a constitutive or stress-inducible promoter is an effective strategy to establish the function of a gene, and another significant strategy is to suppress its expression or knock it out. Significant work regarding this has been initiated in millet. A targeted gene-based functional genomics tool known as transgenomics, or transgenic technology, can be used for gaining valuable information on stress tolerance regulatory mechanisms. But, this technique is still in the initial stages in millets, despite their economic and nutritional importance. However, there are several reports where candidate genes from millet have been functionally validated in *Arabidopsis*, or tobacco (Ramegowda et al., 2012; Yue, Liu and Yu, 2014), which may be used to control the expression of downstream stress responsive genes and can play an important role in improving stress tolerance in pearl millet.

### 7.3.6.4 Transcriptomics

Genome-wide expression profiling studies serve as powerful tools for identifying candidate genes underlying various biological processes and stress regulatory networks (Reddy et al., 2012). A transcriptome may be defined as the total population of mRNAs in a cell, tissue or organism and the analysis related to it is termed "transcriptomics". A comprehensive transcriptome has been developed in pearl millet by combining different independent transcriptome data (Rajaram et al., 2013; Zeng, Conner and Ozias-Akins, 2011). The transcriptome analysis of pearl millet under abiotic stress conditions can provide new insights into gene regulatory networks operating in this crop. Stress regulated pathways in pearl millet can be studied through the detection and characterization of stress responsive genes via transcriptomics and then different approaches can be followed for improving stress tolerance/resistance in millet (Mishra et al., 2007). Several differentially regulated transcripts have been detected for salinity, drought and cold stress, explaining the involvement of a complex gene regulatory network. Recently, high grain zinc development was reported using transgenic expression of zinc transporters, and transcriptomic studies revealed several calcium sensor genes involved in the translocation mechanism in finger millet. Similarly, molecular mechanisms of Fe and Zn homeostasis in soils were studied in *A. thaliana* using transcriptomic and metabolomic approaches (Azevedo et al., 2016), which can be specifically applied to promote further improvement in pearl millet biofortification.

### 7.3.6.5 Proteomics

The term "proteome" refers to the entire set of proteins produced or modified by an organism, and proteomics can be defined as large-scale analysis of proteins expressed under a given condition or stage in a cell, tissue or organism in order to recognize the different regulatory mechanisms. A proteome is a link between its transcriptome and metabolome that is used to study the existent state of any organism. Proteins directly affect several biochemical processes and, thus, proteomics is essential in evaluating plant stress responses, which can also be helpful for understanding different stress regulatory mechanisms and developing stress tolerant varieties in pearl millet [proteome analyses can be done from particular tissues or subcellular fractions by means of the well-established two-dimensional electrophoresis technique and several new approaches developed recently (Barkla et al., 2013)].

### 7.3.6.6 Metabolomics

Genomics, transcriptomics or proteomics have certainly played vital role in understanding genotype and complex biological processes but still are insufficient in predicting phenotype, which is ultimately determined by a cell's metabolite. Consequently, metabolomics is one of the newest and very important techniques among all other omics sciences. It refers to the complete set of low molecular weight compounds of a sample, such as substrates or by-products of enzymatic reactions that directly affect the phenotype of a cell/tissue/organism. Metabolomics determines the profile of a sample's metabolites at a specific time or stage or environmental condition and, thus, gives information about an organism's physiological condition and defines the significance of metabolomics in the functional genomics era (Sumner, Mendes and Dixon, 2003). Metabolite profiling, target analysis and metabolite fingerprinting are some methods of metabolomics which are used for several applications. Thus, an integrated analysis of transcriptome, proteome and metabolome is necessary to completely understand gene function and the molecular events regulating complex biological processes. Metabolomics research has been mainly carried out in model plants and staple cereals, while there is no report of metabolic profiling in pearl millet. There is only one report of metabolic profiling in three varieties of proso millet where gas-chromatography-time-of-flight mass spectrometry (GC-TOFMS) was used to resolve diversity among metabolites (Kim et al., 2013). Thus, this omics approach is also very useful and should be used in pearl millet to explore various aspects of nutricereal.

### 7.3.7 NEXT-GENERATION SEQUENCING (NGS)

Discovery of the NGS techniques helped to increase the pace of millet genome sequencing. These are low cost and high throughput technologies to sequence crops and, thus, helped the development of new cultivars with specific traits. NGS technologies have been utilized to study genomic variations and genetic mapping in many crops (Huang and Han, 2014; Jiao et al., 2012; Zhou et al., 2015). The technological advances in NGS also helped to discover second-generation multi-parent populations, including the nested association mapping population (NAM) and multi-parent advanced generation intercross (MAGIC) population (Bohra, 2013; Davey et al., 2011). It helped to utilize multi-parent allelic diversity for fine QTL mapping for specific traits. Thus, using second-generation populations along with faster bioinformatics tools opens new opportunities to exploit the benefits of both linkage and association mapping (Bevan and Uauy, 2013). NAM facilitates high resolution mapping and identified functional markers that provided information about the genetics involved for complex quantitative traits and variability in different crops (Andersen and Lübberstedt, 2003). On the other hand, the MAGIC population was projected for genome-wide association mapping to identify major genes, QTLs and novel loci linked with chief traits (Bandillo et al., 2013). MAGIC reconstructs the genome of each line developed and helps in dissecting multifarious traits and effective QTL mapping for MAS. Thus, it will help in understanding larger genetic variations in the pearl millet germplasm in comparison to association analysis in naturally occurring haplotypes and, therefore, new alleles for vital traits could be detected by using the multi-parent population (Rakshit, Rakshit and Patil, 2012). However, MAGIC needs more time and resources to establish and illustrate wider segregation, which limits its role in the analysis of complex traits. To study wider genetic variations, PMiGAP was developed at ICRISAT that revealed six subpopulations corresponding to pedigree (Sehgal et al., 2015). It showed the significance of including diverse morphological forms for QTL detection, genome-wide marker discovery and estimation of breeding values using genomics. Thus, it will help to determine new allelic variants within and between races, giving new dimensions for the use of genomic tools for MAS. Sequence data are rapidly accumulating and it is cumbersome to handle the huge amount of generated data, so there is a need to access and exploit the data efficiently by taking advantage of different powerful and friendly databases (Yuan et al., 2001). Thus, invention of low cost and high throughput NGS technology has made it easier and faster to sequence the crops with lower economic value for the development of elite cultivars with desirable traits.

### 7.3.8 GENOTYPING-BY-SEQUENCING (GBS)

Genetic mapping, QTL identification and marker–trait association studies are widely affected by marker density and its uniform distribution

(Pedraza-Garcia et al., 2010). In this context, GBS is an effective and useful technique for crop improvement. It is a whole-genome marker profiling, low-cost technique that is used most widely for marker discovery as well as genotyping. It was used in several crops for constructing libraries for NGS (Elshire et al., 2011). Restriction site-associated DNA sequencing (RAD-seq) and GBS are reduced representation sequencing methods to genotype large populations which involve genomic DNA digestion, ligation with unique barcoded adapters and sequencing of pooled libraries (Elshire et al., 2011; Poland et al., 2012). Several crops, such as wheat, barley, maize, switchgrass, sorghum, and soybean, have taken advantage of a GBS platform to genotype diverse germplasm sets and assist genome-wide association studies (Jarquín et al., 2014; Lu et al., 2013; Poland et al., 2012; Romay et al., 2013). Recently, GBS was used to construct a high-density genetic map with more uniform distribution of markers in pearl millet from a biparental population in comparison to maps constructed earlier (Moumouni et al., 2015). In an another significant study, molecular diversity and population organization analysis of 500 pearl millet accessions was carried out using GBS and 83,875 SNPs were identified, revealing the significance of GBS in genetic diversity assessment (Hu et al., 2015). Thus, GBS can provide a high density SNP detection platform at a cheaper rate, even in the absence of a reference genome sequence, and helps in genotyping large sets of diverse individuals in an effective manner (Elshire et al., 2011).

### 7.3.9 GENETIC MODIFICATIONS/ENGINEERING

Genome, or gene, editing includes several new techniques such as meganucleases (homing endonuclease), zinc finger nucleases (ZFNs), transcriptional activator-like effector nucleases (TALENs) and the CRISPR/Cas nuclease system (clustered regularly interspaced short palindromic repeat/CRISPR-associated protein) but, due to ease of use, precision and cost-efficiency, CRISPR is mostly preferred. The development of CRISPR involves enzyme *Cas9* which is based on the guide RNA molecule to mark a specific DNA sequence and later edits target DNA by making genome alterations either by disrupting genes or inserting new sequences (Ledford, 2015). It is highly efficient, robust, less risky and has a wide variety of agricultural applications. This technique has been applied in several crops, such as rice, *Arabadopsis*, tomato and sorghum (Jiang et al., 2013; Zhou et al., 2014) and can also benefit pearl millet as studies in other cereals suggest that the Cas 9/sg RNA system is very reliable, functional and proficient in a two-model system and, thus, gives plenty of scope for gene manipulations. Genetic modifications can help in nutritional improvement of pearl millet grains and, thus, can be a major research area for nutritional security. Less research on gene editing has been done in pearl millet so far, but good work in major cereals has proven that CRISPR can be a very useful and efficient technique for crop improvement (O'Kennedy, Grootboom and Shewry, 2006). It can be used in pearl millet once Zn and Fe pathways are better understood and can ultimately lead to development of nutrient-rich improved varieties that can offer acceptability to consumers and feasibility to smallholder farmers.

## 7.4 CONCLUSION

Germplasm conservation, cultivar development and QTL mapping, which ultimately benefit genomics-assisted breeding, are some of the factors that are dependent on knowledge of the genomic diversity of a crop. These days, cost effective and high throughput genotyping platforms can be successfully applied to the diploid pearl millet genome. An integrated approach, including transgenics, modern genomics and statistical tools, is required which can ultimately accelerate the genetic gains by increasing yield and its stability under stress. In addition, genetic improvement of pearl millet may be possible through the use of genetic and bioinformatic information, along with precision phenotyping. Further, optimum regeneration techniques targeting the rescuing of the progenies of crosses between economically important millets and their wild relatives are required to introduce important agronomic traits to the cultivated species. A large scale screening methodology has to be developed in order to determine the regeneration capacity for diverse genotypes of millets, as has been investigated for rice. Hence, in future, a robust transformation protocol should be developed for each type and ecotype of pearl millet using the *Agrobacterium* method. On the other hand, the availability of more ESTs and huge genome sequence data need to be utilized in a very effective way to improve pearl millet. The genomics and breeding platform need a better alignment and constant upgrading to develop improved hybrid parental lines and populations must be adapted specifically according to the global agroecologies. Although the successful use of genomic tools, screening and the development of improved genotypes has become easy and fast, and progress towards enhancing the use of genomic resources and NGS in pearl millet is quite useful, still much is awaited in order to execute genomics to improve the nutritional quality of the crop. On the other hand, use of advanced multi-parental and association mapping panels, along with other genomic tools, can speed up the recognition of allelic variants. The construction of high density maps, QTL detection, candidate gene identification, new genome sequence techniques and advanced marker-assisted selection tools are some of the areas which need to be strengthened for pearl millet improvement. The mapping of several abiotic stresses, QTLs and gene pyramiding in pearl millet crop still need to be focused and the recently developed technologies must be tested under actual conditions. The outcomes from model crops can be used in pearl millet to achieve added improvement and develop Zn- and Fe-enriched biofortified varieties. Synteny studies can help to identify common genes associated with nutrition biosynthesis pathways, which may be introgressed into pearl millet by traditional breeding or transgene techniques. Thus, current advances in omics technologies, along with advanced transgenic technology and MAS, will prove useful in targeting pearl millet for genetic improvement to meet the present changing climatic scenario. In conclusion, joint efforts from different areas, such as plant breeding, genomics, bioinformatics, biotechnology, nutrition and genetics etc., are required to move the research and importance of nutricereal pearl millet in a more meaningful direction, thus making the end-product of biotechnological applications safe, more reliable and consumer friendly.

## REFERENCES

Ahuja, I., R. C. De Vos, A. M. Bones, R. D. Hall 2010. Plant molecular stress responses face climate change. *Trends in Plant Science* 15(12):664–674.

Allouis, S., X. Qi, S. Lindup, M. D. Gale, K. M. Devos 2001. Construction of a BAC library of pearl millet, *Pennisetum glaucum*. *Theoretical and Applied Genetics* 102 (8):1200–1205.

Ambawat, S., S. Senthilvel, C. T. Hash, T. Nepolean, V. Rajaram, K. Eshwar, R. Sharma, R. P. Thakur, V. P. Rao, R. C. Yadav, R. K. Srivastava 2016. QTL mapping of pearl millet rust resistance using an integrated DArT-and SSR-based linkage map. *Euphytica* 209(2):461–476.

Andersen, J. R., T. Lübberstedt 2003. Functional markers in plants. *Trends in Plant Science* 8:554–560.

Azevedo, H., S. G. Azinheiro, A. Muñoz-Mérida, P. H. Castro, B. Huettel, M. M. Aarts et al. 2016. Transcriptomic profiling of *Arabidopsis* gene expression in response to varying micronutrient zinc supply. *Genomics Data* 7:256–258.

Bandillo, N., C. Raghavan, P. A. Muyco, M. A. L. Sevilla, I. T. Lobina, C. J. Dilla-Ermita et al. 2013. Multi-parent advanced generation inter-cross (MAGIC) populations in rice: progress and potential for genetics research and breeding. *Rice* 6:11. 10.1186/1939-8433-6-11

Barkla, B. J., T. Castellanos-Cervantes, J. L. D. De Leo, A. Matros, H. P. Mock, F. Perez-Alfocea, G. H. Salekdeh, K. Witzel, C. Zörb 2013. Elucidation of salt stress defense and tolerance mechanisms of crop plants using proteomics—Current achievements and perspectives. *Proteomics* 13:1885–1900.

Bertin, I., J. H. Zhu, M. D. Gale 2005. SSCP-SNP in pearl millet—A new marker system for comparative genetics. *Theoretical and Applied Genetics* 8:1467–1472.

Bevan, M. W., C. Uauy 2013. Genomics reveals new landscapes for crop improvement. *Genome Biology* 14:206.

Bidinger, F. R., T. Nepolean, C. T. Hash, R. S. Yadav, C. J. Howarth 2007. Quantitative trait loci for grain yield in pearl millet under variable post flowering moisture conditions. *Crop Science* 47(3):969–980.

Bohn, M., S. Groh, M. M. Khairallah, D. A. Hoisington, H. F. Utz, A. E. Melchinger 2001. Re-evaluation of the prospects of marker-assisted selection for improving insect resistance against Diatraea spp. in tropical maize by cross validation and independent validation. *Theoretical and Applied Genetics* 103(6–7):1059–1067.

Bohra, A. 2013. Emerging paradigms in genomics-based crop improvement. *Science World Journal* 2013:585467.

Bollam, S., V. Pujarula, R. K. Srivastava, R. Gupta 2018. Genomic approaches to enhance stress tolerance for productivity improvements in pearl millet: genomic approaches. In S. S. Gosal, S. H. Wani (eds), *Biotechnologies of Crop Improvement*. Vol 3. Cham (Vietnam and Cambodia): Springer, Springer International Publishing AG, part of Springer Nature, 239–264. 2018.

Breese, W. A., C. T. Hash, K. M. Devos, C. J. Howarth, J. F. Leslie 2002. Pearl millet genomics and breeding for resistance to downy mildew. In J. F. Leslie (ed), *Sorghum and Millets Diseases*. Ames, Iowa: Iowa State Press, 243–246.

Ceasar, S. A., S. Ignacimuthu 2009. Genetic engineering of millets: current status and future prospects. *Biotechnology Letters* 31:779–788.

Chen, W., W. Wang, M. Peng, L. Gong, Y. Gao, J. Wan et al. 2016. Comparative and parallel genome-wide association studies for metabolic and agronomic traits in cereals. *Nature Communications* 7:12767. doi: 10.1038/ncomms12767

Davey, J. W., P. A. Hohenlohe, P. D. Etter, J. Q. Boone, J. M. Catchen, M. L. Blaxter 2011. Genome-wide genetic marker discovery and genotyping using next-generation sequencing. *Nature Reviews Genetics* 12:499–510.

Deschamps, S., V. Llaca, G. D. May 2012. Genotyping-by-sequencing in plants. *Biology* 1:460–483.

Devos, K. M., M. D. Gale 2000. Genome relationships: the grass model in current research. *Plant Cell* 12(5):637–646.

Devos, K. M., T. S. Pittaway, C. S. Busso, M. D. Gale, J. R. Witcombe, C. T. Hash 1995. Molecular tools for the pearl millet nuclear genome. *International Sorghum and Millets Newsletter* 36:64–66.

Dita, M. A., N. Rispail, E. Prats, D. Rubiales, K. B. Singh 2006. Biotechnology approaches to overcome biotic and abiotic stress constraints in legumes. *Euphytica* 147(1–2):1–24.

Directorate of Millet Development. 2018. Directorate of Millet Development, 710, Mini Secretariate, Bani Park, Jaipur-3022016.

Elshire, R. J., J. C. Glaubitz, Q. Sun, J. A. Poland, K. Kawamoto, E. S. Buckler et al. 2011. A robust, simple genotyping-by-sequencing (GBS) approach for high diversity species. *PLoS One* 6:e19379.

Gale, M. D., K. M. Devos, J. H. Zhu, S. Allouis, M. S. Couchman, H. Liu et al. 2005. New molecular marker technologies for pearl millet improvement. *SAT eJournal* 1:1–7.

Girgi, M., M. M. O'Kennedy, A. Morgenstern, G. Mayer, H. Lorz, K. H. Oldach 2002. Transgenic and herbicide resistant pearl millet (*Pennisetum glaucum* L.) R. Br. via microprojectile bombardment of scutellar tissue. *Molecular Breeding* 10:243–252.

Gurmu, F., H. Mohammed, G. Alemaw 2009. Genotype xenvironment interactions and stability of soybean for grain yield and nutrition quality. *African Crop Science Journal* 17:87–99.

Hash, C. T., R. P. Thakur, V. P. Rao, A. B. Raj 2006. Evidence for enhanced resistance to diverse isolates of pearl millet downy mildew through gene pyramiding. *International Sorghum and Millets Newsletter* 47:134–138.

Hash, C. T. A., J. R. Witcombe 2001. Pearl millet molecular marker research. *International Sorghum and Millets Newsletter* 42:8–15.

Hu, Z., B. Mbacké, R. Perumal, M. C. Guèye, O. Sy, S. Bouchet et al. 2015. Population genomics of pearl millet (*Pennisetum glaucum* (L.) R. Br.): comparative analysis of global accessions and Senegalese landraces. *BMC Genomics* 16:1048.

Huang, X., B. Han 2014. Natural variations and genome-wide association studies in crop plants. *Annual Review of Plant Biology* 65:531–551.

Hyten, D. L., D. J. Lee 2016. Plant genetic mapping techniques. *eLS* 1–8. doi:10.1002/9780470015902.a0002019.pub2

Ignacimuthu, S., P. Kannan 2013. *Agrobacterium* mediated transformation of pearl millet (*Pennisetum typhoides* (L.) R. Br.) for fungal resistance. *Asian Journal of Plant Science* 12:97–108.

Jarquín, D., K. Kocak, L. Posadas, K. Hyma, J. Jedlicka, G. Graef et al. 2014. Genotyping by sequencing for genomic prediction in a soybean breeding population. *BMC Genomics* 15:740.

Jiang, W., H. Zhou, H. Bi, M. Fromm, B. Yang, D. P. Weeks 2013. Demonstration of CRISPR/Cas9/sgRNA-mediated targeted gene modification in *Arabidopsis*, tobacco, sorghum and rice. *Nucleic Acids Research* 41:e188.

Jiao, Y., H. Zhao, L. Ren, W. Song, B. Zeng, J. Guo et al. 2012. Genome-wide genetic changes during modern breeding of maize. *Nature Genetics* 44:812–815.

Jing, H. C., C. Bayon, K. Kanyuka, S. Berry, P. Wenzl, E. Huttner, A. Kilian, K. E. Hammond Kosack 2009. DArT markers: diversity analyses, genomes comparison, mapping and integration with SSR markers in *Triticum monococcum*. *BMC Genome* 10:458.

Jogaiah, S., R. G. Sharath Chandra, N. Raj, A. B. Vedamurthy, H. S. Shetty 2014. Development of SCAR marker associated with downy mildew disease resistance in pearl millet (*Pennisetum glaucum* L.). *Molecular Biology Reports* 41(12):7815–7824.

Jones, E. S., W. A. Breese, C. J. Liu, S. D. Singh, D. S. Shaw, J. R. Witcombe 2002. Mapping quantitative trait loci for resistance to downy mildew in pearl millet. *Crop Science* 42(4):1316–1323.

Kannan, B., S. Senapathy, A. G. Bhasker Raj, S. Chandra, A. Muthiah, A. P. Dhanapal et al. 2014. Association analysis of SSR markers with phenology, grain, and stover-yield related traits in pearl millet (*Pennisetum glaucum* (L.) R. Br.). *The Scientific World Journal* 2014: 562327, 1–14.

Kholová, J., T. Nepolean, C. T. Hash, A. Supriya, V. Rajaram, S. Senthilvel, A. Kakkera, R. Yadav, V. Vadez 2012. Water saving traits co-map with a major terminal drought tolerance quantitative trait locus in pearl millet [*Pennisetum glaucum* (L.) R. Br.]. *Molecular Breeding* 30(3):1337–1353.

Kim, J. K., S. Y. Park, Y. Yeo, H. S. Cho, Y. B. Kim, H. Bae, C. H. Park, J. H. Lee, S. U. Park 2013. Metabolic profiling of millet (*Panicum miliaceum*) using gas chromatography–time-of flight mass spectrometry (GC-TOFMS) for quality assessment. *Plant Omics Journal* 6:73–78.

Kothari, S. L., N. Chandra 1995. Advances in tissue culture and genetic transformation of cereals. *Journal of the Indian Botanical Society* 74:323–342.

Kumari, B. R., M. A. Kolesnikova-Allen, C. T. Hash, S. Senthilvel, T. Nepolean, P. B. Kavi Kishor, O. Riera-Lizarazu, J. R. Witcombe, R. K. Srivastava 2014. Development of a set of chromosome segment substitution lines in pearl millet [*Pennisetum glaucum* (L.) R. Br.]. *Crop Science* 54(6):2175–2182.

Lambé, P., M. Dinani,R. Deltour 2000. Transgenic Pearl millet (*Pennisetum glaucum*). In Y. P. S. Bajaj (ed), Biotechnology in Agriculture and Forestry, Transgenic Crops-I, Vol. 46. Berlin: Springer, 84–108.

Langridge, P., D. Fleury 2011. Making the most of 'omics' for crop breeding. *Trends in Biotechnology* 29(1):33–40.

Lata, C. 2015. Advances in omics for enhancing abiotic stress tolerance in millets. *Procedings of the Indian National Science Academy* 81(2):397–417.

Ledford, H. 2015. CRISPR, the disruptor. *Nature* 522:20–24.

Liu, C. J., J. R. Witcombe, T. S. Pittaway, M. Nash, C. T. Hash, C. S. Busso, M. D. Gale 1994. An RFLP-based genetic map of pearl millet *Pennisetum glaucum*. *Theoretical and Applied Genetics* 89(4):481–487.

Lu, F., A. E. Lipka, J. Glaubitz, R. Elshire, J. H. Cherney, M. D. Casler et al. 2013. Switchgrass genomic diversity, ploidy, and evolution: novel insights from a network-based SNP discovery protocol. *PLoS Genetics* 9:e1003215.

Mace, E. S., L. Xia, D. R. Jordan, K. Halloran, D. K. Parh, E. Huttner, P. Wenzl, A. Kilian 2008. DArT markers: diversity analyses and mapping in Sorghum bicolor. *BMC Genomics* 9:26.

Metzker, M. L. 2010. Sequencing technologies-the next generation. *Nature Reviews Genetics* 11:31–46. 10.1038/nrg2626

Mishra, R. N., P. S. Reddy, S. Nair, G. Markandeya, A. R. Reddy, S. K. Sopory., M. K. Reddy 2007. Isolation and characterization of expressed sequence tags

(ESTs) from subtracted cDNA libraries of *Pennisetum glaucum* seedlings. *Plant Molecular Biology* 64:713–732.

Moghaddam, M. J., S. S. Pourdad 2009. Comparison of parametric and non-parametric methods for analyzing genotype x environment interactions in safflower (*Carthamus tinctorius* L.). *Journal of Agricultural Science* 147:601.

Morris, M., K. Dreher, J. Ribaut, M. Khairallah 2003. Money matters (II): costs of maize inbred line conversion schemes at CIMMYT using conventional and marker-assisted selection. *Molecular Breeding* 11(3):235–247.

Moumouni, K. H., B. A. Kountche, M. Jean, C. T. Hash, Y. Vigouroux, B. G. Haussmann et al. 2015. Construction of a genetic map for pearl millet, *Pennisetum glaucum* (L.) R. Br., using a genotyping-by-sequencing (GBS) approach. *Molecular Breeding* 35:1–10.

Muthamilarasan, M., A. Dhaka, R. Yadav, M. Prasad 2016. Exploration of millet models for developing nutrient rich graminaceous crops. *Plant Science* 242:89–97.

O'Kennedy, M. M., A. Grootboom, P. R. Shewry 2006. Harnessing sorghum and millet biotechnology for food and health. *Journal of Cereal Science* 44:224–235.

Pedraza-Garcia, F., Specht, J.E., and Dweikat, I. (2010). A new PCR-based linkage map in pearl millet. *Crop Science* 50: 1754–1760. doi: 10.2135/cropsci2009.10.0560

Poland, J., J. Endelman, J. Dawson, J. Rutkoski, S. Wu, Y. Manes, S. Dreisigacker, J. Crossa, H. Sánchez-Villeda, M. Sorrells, J. L. Jannink 2012. Genomic selection in wheat breeding using genotyping-by-sequencing. *Plant Genome* 2012 (5):103.

Potrykus, I. 1990. Gene transfer to cereals: an assessment. *BioTechnology* 8:535–542.

Qi, X., T. S. Pittaway, S. Lindup, H. Liu, E. Waterman, F. K. Padi, C. T. Hash, J. Zhu, M. D. Gale, K. M. Devos 2004. An integrated genetic map and a new set of simple sequence repeat markers for pearl millet. *Pennisetum glaucum. Theoretical and Applied Genetics* 109(7):1485–1493.

Rai, K. N., K. Anand Kumar 1994. Pearl millet improvement at ICRISAT-an update. *International Sorghum and Millets Newsletter* 35:1–29.

Rajaram, V., T. Nepolean, S. Senthilvel, R. K. Varshney, V. Vadez, R. K. Srivastava, T. M. Shah, A. Supriya, S. Kumar, B. R. Kumari, A. Bhanuprakash 2013. Pearl millet [*Pennisetum glaucum* (L.) R. Br.] consensus linkage map constructed using four RIL mapping populations and newly developed EST-SSRs. *BMC Genomics* 14(1):159.

Rakshit, S., A. Rakshit, J. V. Patil 2012. Multiparent intercross populations in analysis of quantitative traits. *Journal of Genetics* 91:111–117.

Ramegowda, V., M. Senthil Kumar, K. N. Nataraja, M. K. Reddy, K. S. Mysore, M. Udayakumar 2012. Expression of a finger millet transcription factor, EcNAC1, in tobacco confers abiotic stress-tolerance. *PLoS One* 7:e40397.

Ramya, A. R., M. L. Ahamed, R. K. Srivastava 2017. Genetic diversity analysis among inbred lines of pearl millet [*Pennisetum glaucum* (L.) R. Br.] based on grain yield and yield component characters. *International Journal of Current Microbiology and Applied Science* 6(6):2240–2250.

Reddy, D. S., P. Bhatnagar Mathur, V. Vadez, K. K. Sharma 2012. Grain legumes (Soybean, Chickpea, and Peanut): omics approaches to enhance abiotic stress tolerance. In N. Tuteja, S. S. Gill, A. F. Tiburcio, R. Tuteja (eds), *Improving Crop Resistance to Abiotic Stress*. Weinheim, Germany: Wiley-VCH Verlag GmbH & Co. KGaA. doi: 10.1002/9783527632930.ch39

Repellin, A., M. Baga, P. P. Jauhar, R. N. Chibbar 2001. Genetic enrichment of cereal crops via alien gene transfer: new challenges. *Plant Cell Tissue and Organ Culture* 64:159–183.

Romay, M. C., M. J. Millard, J. C. Glaubitz, J. A. Peiffer, K. L. Swarts, T. M. Casstevens et al. 2013. Comprehensive genotyping of the USA national maize inbred seed bank. *Genome Biology* 14:55.

Sehgal, D., V. Rajaram, I. P. Armstead, V. Vadez, Y. P. Yadav, C. T. Hash, R. S. Yadav 2012. Integration of gene-based markers in a pearl millet genetic map for identification of candidate genes underlying drought tolerance quantitative trait loci. *BMC Plant Biology* 12(1):9.

Sehgal, D., L. Skot, R. Singh, R. K. Srivastava, S. P. Das, J. Taunk, P. C. Sharma, R. Pal, B. Raj, C. T. Hash, R. S. Yadav 2015. Exploring potential of pearl millet germplasm association panel for association mapping of drought tolerance traits. *PLoS One* 10(5):e0122165.

Senthilvel, S., B. Jayashree, V. Mahalakshmi, P. S. Kumar, S. Nakka, T. Nepolean, C. T. Hash 2008. Development and mapping of simple sequence repeat markers for pearl millet from data mining of expressed sequence tags. *BMC Plant Biology* 8(1):119.

Senthilvel, S., T. Nepolean, A. Supriya, V. Rajaram, S. Kumar, C. T. Hash et al. 2010. Development of a molecular linkage map of pearl millet integrating DArT and SSR markers. In *Proceedings of the plant and animal genome 18 conference.* San Diego, CA, 9–13.

Serraj, R., C. T. Hash, S. M. Rizvi, A. Sharma, R. S. Yadav, F. R. Bidinger 2005. Recent advances in marker-assisted selection for drought tolerance in pearl millet. *Plant Production Science* 8(3):334–337.

Singh, P., K. N. Rai, J. R. Witcombe, D. J. Andrews 1988. Population breeding methods in pearl millet improvement (Pennisetum americanum). *Agronomy Tropical* 43:185–193.

Smith, R. L., M. F. Grando, Y. Y. Li, J. C. Seib, R. G. Shatters 2002. Transformation of bahiagrass (*Paspalum notatum flugge*). *Plant Cell Reports* 20:1017–1021.

Strauss, S. H. 2003. Genomics, genetic engineering, and domestication of crops. *Science* 300:61–62.

Sumner, L. W., P. Mendes, R. A. Dixon 2003. Plant metabolomics: large scale phytochemistry in the functional genomics era. *Phytochemistry* 62:817–836.

Supriya, A., S. Senthilvel, T. Nepolean, K. Eshwar, V. Rajaram, R. Shaw, C. T. Hash, A. Kilian, R. C. Yadav, M. L. Narasu 2011. Development of a molecular linkage map of pearl millet integrating DArT and SSR markers. *Theoretical and Applied Genetics* 123:239–250.

Taylor, M. G., V. Vasil, I. K. Vasil 1991. Histology of and physical factors affecting, transient GUS expression in pearl millet (*Pennisetum glaucum* (L.) R. Br.) embryos following microprojectile bombardment. *Plant Cell Reports* 10:120–125.

Vadez, V., T. Hash, J. Kholova 2012. Phenotyping pearl millet for adaptation to drought. *Frontiers in Physiology* 3:386.

Varshney, R. K., T. J. Close, N. K. Singh, D. A. Hoisington, D. R. Cook 2009. Orphan legume crops enter the genomics era. *Current Opinion in Plant Biology* 12 (2):202–210.

Varshney, R. K., D. A. Hoisington, A. K. Tyagi 2006. Advances in cereal genomics and applications in crop breeding. *Trends in Biotechnology* 24(11):490–499.

Varshney, R. K., C. Shi, M. Thudi, C. Mariac, J. Wallace, P. Qi, H. Zhang, Y. Zhao, X. Wang, A. Rathore, R. K. Srivastava 2017. Pearl millet genome sequence provides a resource to improve agronomic traits in arid environments. *Nature Biotechnology* 35(10):969.

Vasil, V., I. K. Vasil 1981. Somatic embryogenesis and plant regeneration from suspension cultures of Pearl millet (*Pennisetum americanum*). *Annals in Botany* 47:669–678.

Velu, G., K. N. Rai, V. Muralidharan, V. N. Kulkarni, T. Longvah, T. S. Raveendran 2007. Prospects of breeding biofortified pearl millet with high grain iron and zinc content. *Plant Breeding* 126(2):182–185.

Wu, J., L. T. Li, M. Li, M. A. Khan, X. G. Li, H. Chen et al. 2014. High-density genetic linkage map construction and identification of fruit-related QTLs in pearl using SNP and SSR markers. *Journal of Experimental Botany* 6:5771–5781.

Yadav, O. P., K. N. Rai 2013. Genetic Improvement of pearl millet in India. *Agricultural Research* 2(4):275–292.

Yadav, R., F. Bidinger, C. Hash, Y. Yadav, O. Yadav, S. Bhatnagar, C. Howarth 2003. Mapping and characterisation of QTL× E interactions for traits determining grain and stover yield in pearl millet. *Theoretical and Applied Genetics* 106 (3):512–520.

Yadav, R. S., C. T. Hash, F. R. Bidinger, K. M. Devos, C. J. Howarth 2004. Genomic regions associated with grain yield and aspects of post-flowering drought tolerance in pearl millet across stress environments and tester background. *Euphytica* 136 (3):265–277.

Yadav, R. S., D. Sehgal, V. Vadez 2011. Using genetic mapping and genomics approaches in understanding and improving drought tolerance in pearl millet. *Journal of Experimental Botany* 62(2):397–408.

Yang, S. Y., R. K. Saxena, P. L. Kulwal, G. J. Ash, A. Dubey, J. D. Harper, H. D. Upadhyaya, R. Gothalwal, A. Kilian, R. K. Varshney 2011. The first genetic map of pigeon pea based on diversity arrays technology (DArT) markers. *Journal of Genetics* 1:103–109.

Yuan, Q., J. Quackenbush, R. Sultana, M. Pertea, S. L. Salzberg, C. R. Bluell 2001. Rice bioinformatics. Analysis of rice sequence data and leveraging the data to other plant species. *Plant Physiology* 125:1166–1174.

Yue, J., L. C. Liu, J. Yu 2014. A remorin gene SiREM6, the target gene of *SiARDP*, from foxtail millet (*Setaria italica*) promotes high salt tolerance in transgenic Arabidopsis. *PLoS One* 9(6): e100772.

Zaveri, P. P., P. S. Phul, B. S. Dhillon 1989. Observed and predicted genetic gains from single and mutli trait selections using three mating designs in pearl millet. *Plant Breeding* 103:270–277.

Zeng, Y., J. Conner, P. Ozias-Akins 2011. Identification of ovule transcripts from the apospory-specific genomic region ASGR)-carrier chromosome. *BMC Genomics* 12:206. 10.1186/1471-2164-12-206

Zhou, H., B. Liu, D. P. Weeks, M. H. Spalding, B. Yang 2014. Large chromosomal deletions and heritable small genetic changes induced by CRISPR/Cas9 in rice. *Nucleic Acids Research* 42:10903–10914.

Zhou, Z., Y. Jiang, Z. Wang, Z. Gou, J. Lyu, W. Li et al. 2015. Resequencing 302 wild and cultivated accessions identifies genes related to domestication and improvement in soybean. *Nature Biotechnology* 33:408–414.

# 8 Biofortification and Medicinal Value of Pearl Millet Flour

*Shobhit, Priyanka Kajla, Supriya Ambawat, Subaran Singh and Suman*

## CONTENTS

## 8.1 INTRODUCTION

Pearl millet [*Pennisetum glaucum* (L) Br.], well known as *bajra*, is among the major nutritionally important cereal grains, recognized for its versatile food, feed and fodder applications, and it is especially grown in semi-arid agricultural areas (Nambiar et al., 2011). This millet, a hardy cereal crop compared to wheat, rice and maize, is cultivated in areas with comparatively low rainfall due to its capacity for tolerance of constant or intermittent circumstances of drought (Dykes and Rooney, 2007; Hoover et al., 1996). Due to its excellent performance in low fertility agricultural lands, such as semi-arid African and Southeast Asian areas, pearl millet serves as a significant staple cereal grown in areas significantly impacted by micromineral malnutrition. Compared to other major cereals, it has a superb nutritional profile and, therefore, has strong potential to contribute towards food and nutrition safety (Malik, 2015).

Pearl millet is established as a nutricereal owing to the presence of good quality protein, unsaturated fatty acids, carbohydrate, insoluble dietary fiber and minerals, particularly iron and zinc. Along with this, millet serves as a storehouse of antioxidant compounds, particularly flavonoids, carotenoids, polyphenols, omega 3 fatty acids and dietary fiber, which have numerous health benefits (NIN, 2017). Among the cereals and pulses, pearl millet is the most economical resource of energy, proteins, iron, phosphorous and zinc. It is also the richest source of energy compared to staple cereals.

Food nutritional quality is the primary element in maintaining overall human health, as nutritional safety is a driving force for health maintenance to maximize genetic potential. Consequently, nutritional quality should be considered when addressing the issue of deep-rooted food insecurity and malnutrition (Singh and Raghuvanshi, 2012). Pearl millet is accredited as a significant millet in developing nations, helping to combat shortage of food and meeting the increasing population's dietary requirements. Thus, it serves as the only major source of energy and proteins in a bigger section of nutritionally deprived populations. Pearl millet has a superior nutritional profile compared to other important cereals because of its high calories, quality protein, lipids and microminerals. It is the fundamental staple food used by the poorest individuals in impoverished nations. It can be used in porridges, leavened and unleavened breads, and can also be boiled or steamed, or used as an ingredient in alcoholic drinks (Patni and Aggarwal, 2017).

Nutritional insecurity is a significant limitation for developing nations whose population is extremely dependent on coarse cereal grains that are usually deficient in micronutrients. Millets are nutritionally superior, due to quality protein, the presence of essential amino acids, microminerals, dietary fiber and vitamins. The dietary potential of this crop has encouraged scientists around the world to focus on improving millet species for developing nations (Kothari et al., 2005). Additionally, epidemiological studies have also shown that frequent consumption of pearl millet can safeguard against the occurrence of heart-related issues, diabetes, gastrointestinal disorders and a variety of health implications (McKeown et al., 2002). Biofortification of commonly consumed cereal grains has been shown to be a cost-effective strategy to tackle malnutrition/deficiency of micronutrients. Biofortification can also be seen as a fresh instrument for combating micronutrient malnutrition, which is more widely acceptable if accomplished through traditional methods of plant breeding. Biofortification in pearl millet is usually done with two aims: first, to make the crop more nutrient dense and, second, elimination of the antinutritional factors to maximize the bioavailability/digestibility of minerals.

## TABLE 8.1
### Nutritional composition of pearl millet

| Nutrients | | | Amount |
|---|---|---|---|
| Macronutrients | Carbohydrates | Sugars (g) | 57.0–76.0 |
| | | Fiber (g) | 1.2–2.0 |
| | Proteins (g) | | 11.6–14.0 |
| | Fats (g) | | 4.8–5.7 |
| | | | |
| Micronutrients | | | |
| Minerals | Calcium (mg) | | 40.0–42.0 |
| | Phosphorous (mg) | | 296–360 |
| | Iron (mg) | | 8.0–11.0 |
| | Zinc (mg) | | 3.1–6.6 |
| | Sodium (mg) | | 9.2–10.9 |
| | Magnesium (mg) | | 97.0–137.0 |
| Vitamins | | | |
| | Vitamin A (mcg) | | 132.0 |
| | Thiamine (mg) | | 0.21–0.38 |
| | Riboflavin (mg) | | 0.2–0.3 |
| | Pantothenic acid (mg) | | 0.5 |
| | Biotin (µg) | | 0.64 |
| | Niacin (mg) | | 0.86–2.3 |
| | Folic acid (mcg) | | 36.0–45.5 |
| | Vitamin E (mg) | | 23–24 |
| | Vitamin K (µg) | | 2.85 |
| | | | |
| Carotenoids | | | |
| | Leutin(µg) | | 29.69 |
| | Zeaxanthin (µg) | | 9.39 |
| | ⊠-carotene (µg) | | 28.83 |
| | | | |
| Antinutrients | | | |
| | Phytic acid (mg/100g) | | 596–618 |
| | Tannins (mg/100g) | | 225–232 |
| | Polyphenols (mg/100g) | | 403–445 |

Source: NIN(2017); Kulthe et al. (2016); Sade (2009).

## 8.2 NUTRITIONAL PROFILE OF PEARL MILLET

Pearl millet is rich in fiber, minerals, antioxidants and resistant starch (Table 8.1).

### 8.2.1 MACRONUTRIENTS

Carbohydrate content in pearl millet varies from 60–70%. About 20–21.5% amylose makes up the starch of the millet and encompasses a comparatively higher water absorption capacity and swelling index than the other cereal starches (Lestieme et al., 2007). Sucrose, glucose, fructose and raffinose are the sugars found in millet (Gupta and Nagar, 2010). Pearl millet is known to have an adequate amount of dietary fiber and most of it is present in insoluble form (National Institute of Nutrition, 2017).

Pearl millet usually contains protein at 9–13%. Lysine is a pearl millet limiting amino acid. Significant inverse correlation exists between the protein concentration and the protein lysine content of millet. The amino acid profile of pearl millet protein includes most of the essential amino acids, which is comparatively higher than sorghum and maize proteins (Adeola et al., 2005). Owing to its good amino acid profile, pearl millet has higher protein digestibility than other grains (Kalinova and Moudry, 2006).

Pearl millet is well known for its higher fat content among all millets and cereal grains (Taylor, 2004). The fatty acid profile of pearl millet is rich in unsaturated fatty acids, particularly linolenic acid (Adeola and Orban, 2005). The higher calorific value of the millet is attributed to its higher fat content (Hanna et al., 1990).

### 8.2.2 MICRONUTRIENTS

Pearl millet contains significant quantities of minerals, specifically iron, calcium, magnesium and phosphorus (Burton et al., 1992). Total mineral content and mineral profile of pearl millet is superior to the other common cereals and millet. It is also rich in B-complex vitamins, as well as lipid-soluble vitamin E and vitamin A (National Institute of Nutrition, 2017). The presence of these may be significant as antioxidants that can reduce the deterioration of triglycerides.

### 8.2.3 ANTINUTRIENTS

Polyphenols, particularly tannins and phytates, are the major antinutritional components present in pearl millet. These antinutritional components are mainly found in the bran layer. There exists a positive correlation between the presence of antinutrients and in vitro protein digestibility. Minimal processing methods reduce the antinutritional factors to significant levels (Irén Léder, 2004).

## 8.3   QUALITY IMPROVEMENT OF PEARL MILLET FLOUR

The full potential of pearl millet flour, despite its superb nutritional profile, is insufficient, due to its tendency to rancidity. Pearl millet is rich in polyunsaturated fatty acids and, especially, the content of linoleic acid is quite high, thus limiting its shelf stability during storage. Lipases carry out the hydrolysis of long chain unsaturated fatty acids into short chain free fatty acids, resulting in the development of off-flavors and rancidity. However, pearl millet is also known to have two types of enzyme systems—peroxidises (which carry out oxidation) (Reddy et al., 1986) and enymes-polyphenol oxidase (catalysis)—which play a significant role in the degradation of the quality of pearl millet flour. The schematic representation of rancidity development in pearl millet flour is presented in Figure 8.1. Thus, the exposure of pearl millet flour to air, light and high temperature causes oxidative damage of the lipids, resulting in the development of free fatty acids, unwanted flavors and odors and, eventually, reduced shelf life.

The rancidity issue also restricts the utilization of pearl millet flour commercially as packaged shelf-stable flour to be utilized in the food processing industry. Development and recognition of rancidity resistant pearl millet varieties and hybrids must be studied in order to tackle rancidity in pearl millet flour and its value-added products by tailoring creative packaging alternatives in combination with processing treatments. Pearl millet varieties prone to low rancidity can be created through conventional breeding, as well as molecular breeding methods, to solve the rancidity issue in pearl millet.

## 8.4   MEDICINAL VALUE OF PEARL MILLET

Millets are universally accepted as nutricereals as these are rich in fats, energy, protein, insoluble dietary fiber, microminerals, vitamins and polyphenols required for safeguarding human health. Wide prospective health advantages of pearl millet have been reported, such as prevention of cancer and

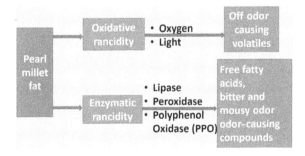

**FIGURE 8.1**   Mechanism of rancidity development in pearl millet flour (**adapted from** Datta Mazumdar et al., 2016)

heart diseases, reduction in the occurrences of tumor, high blood pressure, cholesterol and fat absorption rates, delay in gastric emptying and providing gastrointestinal bulk (Gupta et al., 2012; Truswell, 2002).

Owing to its superb nutritional profile, pearl millet is considered a wonder functional food. The numerous health benefits of pearl millet are listed in Table 8.2. Pearl millet has the lowest glycemic index in comparison to other millets and cereals (Mani et al., 1993), making it appropriate for the management of diabetes.

As pearl millet lacks gluten, the people who are allergic to gluten and suffering from celiac disease can choose it as a nutritious alternative to other cereals. Pearl millet is the only millet having a small prolamin percentage that maintains its alkaline characteristics after cooking, which makes it perfect for individuals with gluten allergy and stomach ulcers. The balanced amino acid profile, along with higher *in vitro* digestibility of protein, makes millet one of the most nutritious and easily digestible option, enriched with quality protein and energy. Pearl millet is known to be heart friendly due to the presence of magnesium. It also increases sensitivity to insulin as well as lowering triglycerides. As the pearl millet is a calorie and nutrient dense source, it can be incorporated in the diets of growing children and pregnant women. Milling of whole pearl millet should be encouraged to make available the maximum benefits of flavonoids and other polyphenolic compounds. Research should be focused more on the development of new food products from pearl millet in order to tailor therapeutic diets for disease management.

## 8.5 BIOFORTIFICATION

Biofortification is defined as the process of improving bioavailability and enhancing the nutrient concentration of crop plants through agronomic practices, plant breeding (White and Broadley, 2005) or modern biotechnology such as recombinant DNA technology (genetic engineering) (Zimmermann and Hurrell, 2002). This process provides a reasonably cost-effective, viable and long-term means of carrying more micronutrients to food crops. However, biofortification differs from traditional fortification, as it aims to raise plant nutrient concentrations during growth compared to manual biofortification during crop processing (Rebecca Bailey, 2008). The advanced biofortification techniques are used when the content of micronutrients cannot be substantially improved through conventional methods. Biofortified staple foods can, thus, help to increase the daily intake of adequate micronutrients in people throughout the lifecycle, offering a viable means to reach malnourished rural populations with restricted access to various commercially fortified foods, supplements and diets (Bouis et al., 2013). Some significant examples of biofortification include iron-biofortification of pearl millet, beans, legumes, rice zinc-biofortification of pearl millet, sweet potato, rice, beans, wheat and cornamino acid, protein biofortification of cassava and sorghum, provitamin A sweet potato, cassava and corn.

## TABLE 8.2

### The medicinal value of bioactive components of pearl millet

| Bioactive component | Health benefits |
|---|---|
| Iron | High iron content improves the haemoglobin levels, thus playing an important role in the cure and prevention of anaemia |
| Lignin, polyphenols and carotenoids | Act as strong antioxidants to reduce the occurrence of cardiovascular disorders |
| Magnesium | Presence of an adequate amount of magnesium helps in maintenance of optimum blood pressure and prevention of heart attack and stroke. Aids in reducing the incidence of respiratory disorders and migraine attacks |
| Phosphorous and calcium | A high amount of phosphorous plays an important role in growth and development of bones |
| High magnesium, phytate, antioxidants | These components reduce tumor development, especially in breast and colon cancer |
| Insoluble dietary fiber | Helps in weight loss (combating obesity) by increasing the satiety value and alleviating constipation |
| Dietary fiber | Total dietary fiber helps in lowering the glycemic index, thus helping in the prevention and cure of diabetes |
| Gluten free | Pearl millet is gluten free, thus serving as a good option for a person suffering from celiac disease (gluten intolerance) |
| Phytic acid | Reduces cholesterol |
| Insoluble fiber | Prevents the formation of gallbladder stones by lowering bile secretion |
| Low prolamine content | As the millet is gluten free, it therefore has a lower prolamine content, which contributes to the alkaline properties and these properties remain even after processing. Thus, it is very useful for the treatment and prevention of stomach ulcers |
| Lactic acid bacteria | Presence of the probiotic aids in diarrhoea treatment |
| Flavonoids, phenols and alpha-linolenic acid | These hamper DNA scission, reducing LDL cholesterol, oxidative stress to membranes and propagation of carcinoma cells; thus, helpful in prevention of non-communicable diseases |
| Antioxidants, phenolics, carotenoids | Anti-ageing properties |
| Chromium | Pearl millet is the only millet to contain adequate levels of chromium Chromium has been reported to be a part of the glucose tolerance factor (GTF) and, therefore, its deficiency leads to impaired glucose tolerance (diabetes) |
| Protein | Good amino acid profile, easily digestible concentrated source of energy, therefore can be best utilized in the formulation of baby foods |

Source: Himanshu et al. (2018); Patni and Aggarwal, (2017); Dayakar Rao et al. (2017).

## 8.6  BIOFORTIFICATION IMPLEMENTATION

The following three questions need to be answered for successful implementation of the biofortification process:

- Can plant breeding be capable of increasing the micronutrient density in staple foods to the levels that will have a quantifiable and major impact on nutritional status?
- Will additional mineral nutrients bred into staple foods be absorbed and utilized sufficiently to increase the status of micronutrients, if consumed under controlled conditions?
- Will farmers grow, and consumers accept, the biofortified varieties in sufficient quantities?An impact pathway with three phases of discovery, development and dissemination, suggested by Saltzman et al. (2013), seeks to address the above questions in the following ways.

---

**Discovery Phase**

- Identification of target populations and setting nutrient targets
- Validation of set nutrient targets
- Discovery and screening of crop genes

---

---

**Development Phase**

- Improving and evaluating the crops
- Testing nutritional efficacy of crops
- Studying farmer adoption and consumer acceptance

---

---

**Dissemination Phase**

- Release and dissemination of crops in target countries
- Increasing consumption of crops

---

Measuring crop adoption and nutritional improvement status
*Pathway depicting biofortification impact and implementation*

## 8.6.1  Discovery

Consumption trends and micronutrient malnutrition are the key factors for identifying target population and focus crops. Considering the usual food consumption patterns and average food intake of target population groups, bioavailability of nutrients and losses of nutrient during processing and storage, the nutritional targets are set by nutritionists in collaboration with breeders (Hotz and McClafferty, 2007). Thereafter, breeders screen available germplasm to discover whether sufficient genetic variation exists for a particular nutrient trait. In addition, high yielding crop varieties which are highly resistant to insect pests and diseases and tolerant to various abiotic stresses may be preferred to biofortify for various micronutrients such as iron, zinc, calcium, essential amino acids and vitamin A. Studies conducted in the past have shown that selection of germplasm having different mineral profiles and vitamins can be used to achieve the desired genetic improvement (Dwivedi et al., 2012; Fageria et al., 2012; Gomez-Becerra et al., 2010; Jiang et al., 2008; Kumar et al., 2012; Menkir, 2008; Menkir et al., 2008; Talukder et al., 2010; Velu et al., 2012). Genetic transformation in the suitable population is an appropriate option to incorporate specific genes for desirable nutrient traits to ensure whether the target of nutritional density had been successfully met or not.

## 8.6.2  Development

Crop quality improvement in terms of nutritional content, high agronomic performance and consumer preferred quality is undertaken through various breeding activities. Thereafter, a high yielding and high nutrient crop variety is released for cultivation. Breeding, testing and release of varieties is a lengthy process, taking 6–10 years. The nutritional efficacy of crops is measured on the basis of retention of micronutrients and their bioavailability in the target crop under various processing, storage and cooking practices. Last, economic research on the assessment of biofortified varieties for consumer acceptance and their adoption by farmers, informs crop improvement researchers during the development phase.

## 8.6.3  Dissemination

Before delivering biofortified crops to the target populations, these should be released for cultivation in the target countries. For maximizing adoption and promoting consumption of these crops, economists undertake studies on the production of seed and grains in order to report on efficient, targeted delivery and marketing strategies, keeping in mind consumer acceptance and varietal adoption. Last, nutritionists ensure improvement in the nutritional status of these crops.

## 8.7 BIOFORTIFICATION APPROACHES

Biofortification was considered as a long-lasting strategy to improve the mineral composition of crops at source (Zhu et al. 2007). It involves increasing mineral levels as well as the bioavailability in grains of staple crops. Plant breeding and genetic engineering involve changing the genotype of a target crop by developing germplasm lines with genes favoring the most efficient build-up of bioavailable minerals. This could be achieved by undertaking a crossing program of the best performing crop plants and selecting only those with desirable traits. Genetic engineering involves the introduction of genes accessed from any source directly into the crop plants. The main advantage of these approaches for improvement in mineral content lies in the fact that the investment is essentially required for research and development activities. In addition, mineral-rich plants tend to be more vigorous and tolerant of biotic stresses, thus leading to improved yield in line with mineral content (Frossard et al., 2000; Nestel et al., 2006). Plant breeding and genetic engineering, which are better and more sustainable approaches, are required to achieve the desired targets (Stein et al., 2008). Although presently very few commercial plants are available with better nutrition derived from these methods, yet these approaches are cost-effective and could have a vital impact in future. Biofortification will be more accessible in the long term because it eliminates hurdles and is independent of any infrastructure or procurement. Also, the normal taste and texture of grains remains unchanged, as plants absorb minerals in organic forms, which is obviously bioavailable. Biofortification strategies, in combination with conventional strategies, have numerous potential health benefits (Buois, 2002; Stein et al., 2008).

### 8.7.1 CONVENTIONAL BREEDING

Crop scientists use conventional breeding to considerably improve the nutritional and quality parameters, as well as the agronomic traits, of pearl millet. But this technique is time consuming, taking 8–10 years to transfer a trait from a donor species to a crop cultivar. It has limited application, since it uses accessible and observed genetic variation in the crop to be improved or is sometimes used in the wild varieties that are crossed with the crop. In order to achieve a higher level of nutrition, sometimes the breeders forgo yield or grain quality parameters. This approach requires decades of breeding effort to develop varieties suitable for growers. However, many advantages are possible where the characteristics that lead to more minerals in the plant can also provide higher yield. Biofortified crops have been developed successfully for both nutrient and yield attributes (Unnevehr et al., 2007).

In plant breeding programs, natural genetic variation is used to enhance bioavailability and the mineral level in crop (Welch and Graham, 2005). Breeding strategies focus on unearthing genetic variability influencing heritable mineral traits, examining stability under varied environmental condtions and the possibility for enhancing mineral content in grains without influencing yield and quality characteristics. These strategies for enhancing mineral content have numerous benefits, and the development of biofortified varieties is a lengthy process because mineral traits have to be introduced from wild relatives. These mineral-rich varieties have not been commercialized so far on a large scale. Breeders use molecular biology techniques to speed up the identification of high mineral varieties, but soil properties must also be considered, as these may hamper mineral uptake and the accumulation process. The minerals available to plant roots could be very low in dry, alkaline soils, along with low organic content (Cakmak, 2008).

### 8.7.2 MUTATION BREEDING

This has been used widely for improving grain quality, along with a higher yield and other traits, in crop improvement programs. In this technique, mutations are induced by chemical treatments or irradiation which, in turn, leads to enhanced genetic variability. The mutation breeding technique has been extensively used all over the world; more than 2,500 varieties have been developed and placed on the FAO/IAEA website. Biofortification was included as a major target in a mutagenesis program conducted by FAO/IAEA, but no applicable results have been achieved so far.

### 8.7.3 MOLECULAR BREEDING

Molecular breeding, or marker-assisted breeding, which relies on molecular markers in breeding processes, has become an extremely important biotechnological method and is being widely used in the development of advanced genomic tools. It is a less time consuming and faster process for developing a pure line variety as compared to conventional breeding. Its use has been increased significantly by plant breeders in public and private sectors (Pray, 2006). Moreover, using this technique, recessive traits could also be located in plants, which was difficult with conventional breeding. Thus, many different genes, coding for different traits, can be stacked in one variety using this tool (Pray, 2006).

### 8.7.4 GENETIC ENGINEERING

Genetic engineering (GE) is the most recent technique being used to increase minerals in the pearl millet crop and overcome their deficiencies. It involves the use of modern biotechnological techniques for introducing genes directly to pearl millet varieties. It will prove to be an efficient approach for transferring desired traits to a crop. The genes from any source are evaluated for their ability to achieve different goals, such as: (a) improving efficiency of mineral mobilization in the soil, (b) reducing the level of antinutritional compounds, (c) increasing the level of nutritional enhancer (Zhu et al., 2007). Genes with desired traits can be transferred from the source organism to the target organism via GE with greater pace and reach. The transgenics, or

GMOs, are produced by this technique. It adds valuable characteristics that are absent in the seeds of individual plant species and desired traits for a crop plant can be incorporated in a short time. Although conventional breeding is less costly in terms of infrastructure and highly skilled manpower, as well as having fewer regulatory hurdles to overcome, it takes more time, and to obtain the desired benefits faster, GE is the only option. Moreover, it becomes easy to develop varieties having several specific nutritional characteristics, sustaining agronomic viability in biofortified crops through GE.

### 8.7.5   Tissue Culture

This technique offers the opportunity to replicate millions of plants from a single cell and is used widely to produce healthy material for clonally propagated crops. By integrating this technique with embryo rescue, genes from wild relatives, which are usually not crossed with a cultivated variety under normal situations, can be used by breeders to enhance crop improvement. Thus, it helps to increase genetic variation in cultivated crops by incorporating important traits of wild relatives and different traits can be targeted to develop superior varieties with desirable agronomic traits.

### 8.7.6   Microbiological Interventions

The utilization of plant growth promoting rhizobacteria (PGPR) has been progressively rising in agriculture, as it helps in reducing the amount of agrochemicals, pesticides and fertilizers (Rana et al., 2012). Phytosiderophores secreted by microorganisms can be used effectively for uptake of micronutrients by plants from the rhizosphere. PGPR constitutes an important piece of protective flora which helps to enhance root function, suppress disease and accelerate growth and development (Glick, 1995). Azotobacter differed from crop plants in competitiveness for Fe and Zn extraction (Shivay et al., 2010). PGPR might be a good parameter that could increase micronutrient contents in crops and yield, along with breeding.

Mycorrhizal fungi associated with plants are normally helpful in improving mineral uptake from the soil, thus enhancing both plant growth and productivity, as they modify the mineral content of plant products. Thus, mycorrhizas offer an efficient and sustainable element of biofortification to control human malnutrition. Six different kinds of mycorrhizas: arbuscular, arbutoid, ecto-, ericoid, monotropoid and orchid have been identified on the basis of their discrete morphological characteristics (Wang and Qiu, 2006). Out of these, *arbuscular mycorrhiza* (AM) is the most important mycorrhiza used in agriculture. The AM explores soil substrates and acquires major macronutrients such as N, P, K, and micronutrients Cu, Fe, Zn (Caris et al., 1998) with a little less capability for taking organic N and P (Koide and Kabir, 2000). They are essential for both AM development and the host plant. Mycorrhizal mycelia and their exudates help in improving nutrient availability in the soil by providing a carbon source for the proper functioning of other below-ground microorganisms. AM fungi play a vital role in improving soil texture and the health of the whole ecosystem and, thus, improve plant performance.

### 8.7.7 AGRONOMIC INTERVENTIONS

Fertilizers are added to soil up to a certain limit in order to improve the health of plants and to enhance mineral accretion in grains for nutritional benefits (Rengel et al., 1999). The approach is functional only if a lack of mineral in the grain reflects its absence in the soil and applied fertilizers are quickly and easily mobilizable. This agronomic strategy is possibly good and more functional in niches or in conjunction with other approaches (Cakmak, 2008). Its disadvantage is that fertilizer use increases the cost of food, thus reducing its accessibility to the poorest people. Also, increased fertilizer use may have environmental impacts (Graham, 2003). The agronomic approach to augment the content of minerals in edible tissues normally depends upon fertilizer application and/or improved solubilization and mobilization of minerals in the soil (White and Broadley, 2009). In the case of immediate non-availability of minerals to crops, targeted use of soluble inorganic fertilizers to roots is followed, but its foliar applications are more useful where minerals are not translocated to edible tissues. In view of an increasing global population, use of fertilizers is essential to achieve a higher crop yield to meet food demands (Graham et al., 2007). Economic yields are increased by applying nutrients mainly to soil and foliage. Nutrient application on the basis of soil testing is a more effective and common method when required in higher amounts. However, the foliar application of fertilizers is more effective and economic when nutrient deficiency is exhibited by visual symptoms or plant tissue tests.

## 8.8 PEARL MILLET—THE MOST SUITABLE CROP FOR BIOFORTIFICATION

Pearl millet, a highly nutritious multipurpose cereal, is grown on about 27 mha in the world (Jalaja et al., 2016). It thrives well in low moisture, nutrient-depleted soils and at high temperatures (> 40°C) because it has evolved under the harsh climatic conditions of infertile soils, heat and drought. It is largely cultivated under rain-fed conditions of arid to semi-arid parts of Western and Eastern Africa and India (Oumar et al., 2008). In these areas, it is an important part of local diets because its production is 75% of the total cereals (Lestienne et al., 2005). Its grains are comparatively more nutritious than other cereals and generally the crop needs few chemicals, resulting in low production costs, making it a more suitable crop for areas not benefiting from dominant agricultural production systems (Jalaja et al. 2016). It is an orphan crop generally consumed by poor people suffering from micronutrient malnutrition. In non-pregnant, non-lactating women, consumption of 250 g/day of biofortified pearl millet met 84% of the RDA for Fe and 100% of the RDA for Zn, whereas ordinary pearl millet provided only 20% of their Fe needs. It is a rich source of dietary protein, carbohydrate, fat, ash, vitamins and minerals. It is gluten free and retains its alkaline properties after cooking and is ideal for gluten allergic people. Due to its excellent nutritional properties, resilience to climate change and consumption by poor people, it is the most suitable crop for Fe and Zn biofortification. In pearl millet, biofortification can be attained either by enhancing the accumulation of nutrients in milled grains, or reducing the antinutrients to enhance the bioavailability of minerals. In the pearl millet crop, there is genetic material with a very large variability for grain Fe content (31–125 ppm) and Zn content (35–82 ppm). In pearl millet, the global baseline for Fe content has been 47 ppm and the target to reduce Fe deficiencies is 77 ppm (Rai et al., 2013). The ICAR–AICRP on pearl millet set benchmark levels at a minimum 42 ppm of Fe and 32 ppm of Zn as the promotion criteria, based on studies conducted jointly by ICRISAT and AICRP on pearl millet (Rai et al. 2015).

## 8.9  BIOFORTIFIED PEARL MILLET CULTIVARS

Although micronutrients are required in traces, they are absolutely vital for various physiological functions. Their deficiencies are termed as hidden hunger. Globally, around two billion people in underdeveloped and developing countries are suffering from Fe and Zn deficiencies. Severe Fe deficiency causes anemia, which results in maternal and child mortality. Generally, children with Zn deficiency are at risk of pneumonia, stunting, diarrhea and mortality. The breeding of biofortified cultivars with higher micronutrients, particularly Fe and Zn, has been recognized as a cost effective and sustainable approach among various strategies to address micronutrient deficiencies.

A large variability in pearl millet cultivars for Fe and Zn content has been observed at ICRISAT, Hyderabad. In a multi-location trial, Fe and Zn content ranged from 42 mg/kg to 67 mg/kg and 37–52 mg/kg in varieties, whereas, in hybrids, the corresponding values were in the range of 31–61 mg/kg and 32–52 mg/kg, respectively. In India, generally farmers preferred hybrids to open-pollinated varieties of pearl millet due to higher grain yield and greater uniformity for various traits. Thus, a pearl millet biofortification program needs to be focused towards breeding cultivars with higher Fe and Zn content.

ICRISAT bred high Fe content pearl millet varieties ICTP 8203 in 1988 (67 mg/kg) and ICMV 221 in 1993 (61 mg/kg), while, the corresponding Zn contents were 52 mg/kg and 45 mg/kg. Pearl millet hybrids Ajeet 38, Proagro XL 51, PAC 903 and 86M86 have been developed with an Fe content of 55–56 mg/kg and Zn content of 39–41 mg/kg. The improved version of variety ICTP 8203, having an Fe content of 71 mg/kg without any change in Zn content, was released in 2014 as *Dhanashakti*, which has been rapidly adopted by the farmers. Likewise, variety ICMV 221Fe11-2, a better version of variety ICMV 221, has been developed with high Fe (81 mg/kg) and Zn (51 mg/kg) content. Hybrids ICMH 1201 and ICMH 1301 have been developed at

ICRISAT with Fe content of 75 mg/kg and 77 mg/kg, respectively. These hybrids had a yield advantage of 36% and 33% over variety ICTP 8203, respectively (ICRISAT, India). Biofortified pearl millet hybrid HHB 299 was developed by CCSHAU, Hisar with an Fe content of 73 ppm and average grain yield of 39.5 q/ha, which was notified in 2018 (Anonymous, 2018). Also, biofortified hybrid AHB 1200Fe has been notified and four other bio-fortified hybrids, RHB 233, RHB 234, HHB 311 and AHB 1269, have been released during 2018. In a pearl millet improvement program in India, breeding lines with >80 ppm Fe content have been identified and with >90 ppm have been developed. ICRISAT has also identified pearl millet breeding material with 90–100 mg/kg Fe and 70–80 mg/kg Zn content. The development of biofortified varieties and hybrids having higher Fe and Zn content will help to fight their deficiencies.

## 8.10  CONCLUSION

Pearl millet is a highly nutritious crop with good grain quality and a considerable amount of essential amino acids, vitamins and minerals such as Fe and Zn. It is a well-known fact that Fe and Zn deficiencies can cause detrimental effects on health and can lead to malnutrition and hidden hunger. Thus, it is high time to invest in new initiatives for scientific research to utilize the highly nutritious millet crop in order to overcome micronutrient malnutrition. Owing to a lack of nutritious food, many people depend on food supplements and fortification methods to fight malnutrition. But, these strategies are not reliable and have several drawbacks. Thus, biofortification is a comparatively more stable approach that is usually accomplished by traditional plant breeding and transgenic techniques. Although a great deal of effort has been devoted to targeting genes for enhancing Fe and Zn levels in hybrid varieties, still more input is required in this direction to enhance millet improvement. In addition, the bioavailability of nutrients is a major issue that needs to be addressed. Many bioavailability research studies are going on and the studying of levels of bioavailability for Fe and Zn needs further development in future as well. The use of available genetic resources, diverse germplasm collections and genomics tools can play a vital role in this direction. A positive and highly significant correlation between Fe and Zn has been reported, thus providing good opportunities for enhancing levels of both micronutrients simultaneously without affecting yield. Further, different missions such as those looking at nutrition and those investigating millet, supported by the government of India, must work together with agriculture research and production organization to increase the benefits and meet nutrition commitments. The biofortification journey is quite long, and issues are being discussed on several platforms but still much effort is needed to take this aspect into the mainstream. Thus, biofortified pearl millet development programs should be considered a matter of priority, so that they can significantly contribute to improved nutrition in future.

## REFERENCES

Adeola, O. and Orban, J.I. 2005. Chemical composition and nutrient digestibility of pearl millet (*Pennisetum glaucum L.*) fed to growing pigs. *Journal of Cereal Science*. 22: 174–184.

Anonymous. 2018. *Annual Report*. Chaudhary Charan Singh Haryana Agricultural University, Hisar, Haryana.

Bouis, H., Low, J., McEwan, M., and Tanumihardj, S. 2013. Biofortification: Evidence and lessons learned linking agriculture and nutrition. FAO Publication. 23. www.fao.org/publications

Bouis, H.E. 2002. Plant breeding: a new tool for fighting micronutrient malnutrition. *Journal of Nutrition*. 132: 491S–94S.

Burton, G.W., Wallace, A., and Rachie, K.O. 1992. Chemical composition and nutritive value of pearl millet (*Pennisetum typhoides*) grain. *Crop Science*. 12: 187–188.

Cakmak, I. 2008. Enrichment of cereal grains with zinc: agronomic or genetic biofortification. *Plant Soil*. 302: 1–17.

Caris, C., Hordt, W., Hawkins, H.J., Romheld, V. and George, E. 1998. Studies of iron transport by arbuscular mycorrhizal hyphae from soil to peanut and sorghum plants. *Mycorrhiza*. 8: 35–39.

Dayakar Rao, B., Bhaskarachary, K., Arlene Christina, G.D., Sudha Devi, G., Vilas A. Tonapi, 2017. *Nutritional and health benefits of millets. ICAR_Indian Institute of Millets Research (IIMR)* Rajendranagar, Hyderabad, pp: 112.

Datta Mazumdar, S., Gupta, S.K., Banerjee, R., Gite, S., Durgalla, P., and Bagade, P. 2016. Determination of variability in rancidity profile of selected commercial pearl millet varieties/hybrids. Poster presented at CGIAR Research Program on Dryland Cereals Review Meeting ICRISAT Hyderabad, India. http://drylandcereals.cgiar.org/index.php/determination-of-variability-in-rancidity-profile-of-select-commercial-pearl-millet-varietieshybrids/.

Dwivedi, S.L. Sahrawat, K., Rai, K.N., Blair, M.W., Andersson, M.S., and Pfeiffer, W. 2012. Nutritionally enhanced staple food crops. In *Plant Breeding Reviews*, ed. J. Janick, Vol 36. 169–291. John Wiley & Sons, USA.

Dykes, L. and Rooney, L. 2007. Phenolic compounds in cereal grains and their health benefits. *Cereal Foods World*. 52: 105–111.

Fageria, N.K., Moraes, M.F., Ferreira, E.P.B., and Knupp, A.M. 2012. Biofortification of trace elements in food crops for human health. *Communications in Soil Science and Plant Analysis*. 43: 556–570.

Frossard, E., Bucher, M., Machler, F., Mozafar, A., and Hurrell, R. 2000. Potential for increasing the content and bioavailability of Fe, Zn and Ca in plants for human nutrition. *Journal of the Science of. Food and Agriculture*. 80: 861–879.

Glick, B.R. 1995. The enhancement of plant growth by free-living bacteria. *Canadian Journal of Microbiology*. 41: 109–117.

Gomez-Becerra, H.F., Yazici, A., Ozturk, L., Budak, H., Peleg, Z., Morgounov, A., Fahima, T. 2010. Genetic variation and environmental stability of grain mineral nutrient concentrations in *Triticum dicoccoides* under five environments. *Euphytica*. 171: 39–52.

Graham, R.D. 2003. Biofortification: a global challenge program. *International Rice Research Notes*. 28: 4–8.

Graham, R.D., Welch, R.M., Saunders, D.A., Ortiz, Monasterio I., Bouis, H.E., Bonierbale, M., Haan, D.E., Burgos, G., Thiele, G., Liria, R., Meisner, C.A., Beebe, S.E., Potts, M.J., Kadian, M., Hobbs, P.R., Gupta, R.K., and Twomlow, S. 2007. Nutritious subsistence food systems. *Advances in Agronomy*. 92: 1–74.

Gupta, N., Srivastava, A.K., and Pandey, V.N. 2012. Biodiversity and nutraceutical quality of some Indian millets. *Proceedings of the National Academy of Sciences, India Section B: Biological Sciences.* www.springerlink.com/ doi: 10.1007/s40011-012-0035-z.

Gupta, V. and Nagar, R. 2010. Effect of cooking, fermentation, dehulling and utensils on antioxidants present in pearl millet *rabadi* – a traditional fermented food. *Journal of Food Science and Technology.* 47: 73–76.

Hanna, A., Singh, J., Faubion, J.N., and Hossney, R.C. 1990. Studies on odour generation in ground pearl millet. *Cereal Food World.* 5: 838–840.

Himanshu, K., Chauhan, M., Sonawane, S.K., and Arya, S.S. 2018 Nutritional and nutraceutical properties of millets: A review. *Clinical Journal of Nutraceutical and Diatetics* 1: 1–10.

Hoover, R., Swamidas, G., Kok, L.S., and Vasanthan, T. 1996. Composition and pysicochemical properties of starch from pearl millet grains. *Food Chemistry.* 56: 355–367.

Hotz, C. and McClafferty, B. 2007. From harvest to health: challenges for developing biofortified staple foods and determining their impact on micronutrient status. *Food and Nutrition Bulletin.* 28: S–271–79.

Irén Léder. 2004. Sorghum and millets in cultivated plants, primarily as food sources: In *Encyclopedia of Life Support Systems (EOLSS).* ed. György Füleky. Eolss Publishers, Oxford.

Jalaja, N., Maheshwari, P., Naidu, K.R., and Kavi Kishor, P.B. 2016. *In-vitro* regeneration and optimization of conditions for transformation methods in pearl millet, *Pennisetum glaucum* (L.). *International Journal of Clinical and Biological Sciences.* 1: 34–52.

Jiang, S.L., Wu, J.G., Thang, N.B., Feng, Y., Yang, X.E., and Shi, C.H. 2008. Genotypic variation of mineral elements contents in rice (*Oryza sativa* L.). *European Food Research and Technology.* 228: 115–122.

Kalinova, J. and Moudry, J. 2006. Content and quality of protein in proso millet (*Panicum miliaceum L.*) varieties. *Plant Foods Human Nutrition.* 61: 45–49.

Koide, R.T. and Kabir, Z. 2000. Extraradical hyphae of the mycorrhizal fungus *Glomus intraradices* can hydrolyse organic phosphate. *New Phytology.* 148: 511–517.

Kothari, S.L., Kumar, S., Vishnoi, R.K., Kothari, A., and Watanabe, K.N. 2005. Applications of biotechnology for improvement of millet crops: Review of progress and future prospects. *Plant Biotechnology.* 22: 81–88.

Kulthe, A.A., Thorat, S.S., and Lande, S.B. 2016. Characterization of pearl millet cultivars for proximate composition, minerals and anti-nutritional contents. *Advances in Life Sciences.* 5: 4672–4675.

Kumar, A., Reddy, B.V.S., Ramaiah, B., Sahrawat, K.L., and Pfeiffer, W.H. 2012. Genetic variability and character association for grain iron and zinc contents in sorghum germplasm accessions and commercial cultivars. *European Journal of Plant Science and Biotechnology.* 6: 66–70.

Lestienne, I., Besançon, P., Caporiccio, B., Lullien-Péllerin, V., and Tréche, S. 2005. Iron and zinc *in vitro* availability in pearl millet flours (*Pennisetum glaucum*) with varying phytate, tannin, and fiber contents. *Journal Agricultural and Food Chemistry.* 53: 3240–3247.

Lestienne, I., Buisson, M., Lullien-Pellerin, V., Picq, C., and Trèche, S. 2007. Losses of nutrients and anti-nutritional factors during abrasive decortication of two pearl millet cultivars (*Pennisetum glaucum*). *Food Chemistry.* 100: 1316–1323.

Malik, S. 2015. Pearl millet-nutritional value and medicinal uses. *International journal of Advanced Research and Innovative Ideas in Education.* 1: 414–418.

Mani, U.V., Prabhu, B.M., Damle, S.S., and Mani, I. 1993. Glycemic index of some commonly consumed foods in Western India. *Asia Pacific Journal of Clinical Nutrition*. 2: 111–114.

McKeown, N.M., Meigs, J.B., Liu. S., Wilson, P.W., and Jacques, P.F. 2002. Whole-grain intake is favorably associated with metabolic risk factors for type 2 diabetes and cardiovascular disease in the Framingham offspring study. *American Journal of Clinical Nutrition*. 76: 390–398.

Menkir, A. 2008. Genetic variation for grain mineral content in tropical-adapted maize inbred lines. *Food Chemistry*. 110: 454–464.

Menkir, A., Liu, W., White, W., Maziya-Dixon, B., and Rocheford, T. 2008. Carotenoid diversity in tropical-adapted yellow maize inbred lines. *Food Chemistry*. 109: 521–529.

Nambiar, V.S., Dhaduk, J.J., Sareen, N., Shahu, T., and Desai, R. 2011. Potential functional implications of pearl millet (*Pennisetum glaucum*) in health and disease. *Journal of Applied Pharmaceutical Science*. 01: 62–67.

National Institute of Nutrition. 2017. *Nutritive value of Indian foods*. eds. Gopalan, C., and Deosthale Y.G. National Institute of Nutrition, Hyderabad.

Nestel, P., Buois, H.E., Meenakshi, J.V., and Pfeiffer, W. 2006. Biofortification of staple food crops. *Journal of Nutrition*. 136: 1064–1067.

Oumar, I., Mariac, C., Pham, J.L., and Vigouroux, Y. 2008. Phylogeny and origin of pearl millet (*Pennisetum glaucum* [L.] R. Br) as revealed by microsatellite loci. *Theoretical Applied Genetics*. 117: 489–497.

Patni, D. and Agrawal, M. 2017. Wonder millet – pearl millet, nutrient composition and potential health benefits – a review. *International Journal of Innovative Research and Review*. 5: 6–14.

Pray, C. 2006. *The Asian Maize Biotechnology Network (AMBIONET): a model for strengthening national agricultural research systems*. International Maize and Wheat Improvement Center (CIMMYT), Mexico.

Rai, K.N., Velu, G., Govindaraj, M., Upadhyaya, H.D., Rao, A.S., and Shivade, H. 2015. Iniadi pearl millet germplasm as a valuable genetic resource for high grain iron and zinc densities. *Plant Genetic Resources*. 13: 75–82.

Rai, K.N., Yadav, O.P., Rajpurohit, B.S., Patil, H.T., Govindaraj, M., and Khairwal, I.S. 2013. Breeding pearl millet cultivars for high iron density with zinc density as an associated trait. *Journal of Semi-Arid Tropical Agricultural Research*. 11: 1–7.

Rana, A., Joshi, M., Prasanna, R., Shivay, Y.S., and Nain, L. 2012. Biofortification of wheat through inoculation of plant growth promoting rhizobacteria and cyanobacteria. *European Journal of Soil Biology*. 50: 118–126.

Rebecca B. 'Biofortifying' one of the world's primary foods. Retrieved on July 22, 2008.

Reddy, V.P., Faubion, J.M., and Hoseney, R.C. 1986. Odor generation in ground, stored pearl millet. *Cereal Chemistry*. 63: 403–406.

Rengel, Z., Batten, G.D., and Crowley, D.E. 1999. Agronomic approaches for improving the micronutrient density in edible portions of field crops. *Field Crop Research*. 60: 27–40.

Sade, F.O. 2009. Proximate, antinutritional factors and functional properties of processed pearl millet (*Pennisetum glaucum*). *Journal of Food Technology*. 7: 92–97.

Saltzman, A., Birol, E., Bouis, H.E., Boy, E., De Moura, F.F., Islam, Y., and Pfeiffer, W.H. 2013. Biofortification: Progress toward a more nourishing future. *Global Food Security*. 2: 9–17.

Shivay, Y.S., Prasad, R., and Rahal, A. 2010. Studies on some nutritional quality parameters of organically or conventionally grown wheat. *Cereal Research Communication*. 38: 345–352.

Singh, P. and Raghuvanshi, R.S. 2012. Finger millet for food and nutritional security. *African Journal of Food Science*. 6: 77–84.

Stein, A.J., Meenakshi, J.V., Qaim, M., Nestel, P., Sachdev, H.P.S., and Bhutta, Z.A. 2008. Potential impacts of iron biofortification in India. *Social Science and Medicine.* 66: 1797 1808.

Talukder, Z.I., Anderson, E., Miklas, P.N., Blair, M.W., Osorno, J., Dilawari, M., and Hossain, K.G. 2010. Genetic diversity and selection of genotypes to enhance Zn and Fe content in common bean. *Canadian Journal of Plant Science.* 90: 49–60.

Taylor, J.R.N. 2004. Millet|Pearl. *Encyclopedia of Grain Science.* 2: 253–261.

Truswell, A.S. 2002. Cereal grain and coronary heart disease. *European Journal of Clinical Nutrition.* 56: 1–4.

Unnevehr, L., Pray, C., and Paarlberg, R. 2007. Addressing micronutrient deficiencies: alternative interventions and technologies. *AgBioforum-Journal of Agrobiotechnology Management & Economics.* 10: 124–134.

Velu, G., Singh, R.P., Huerta-Espino, J., Pena, R.J., Arun, B., Mahendru-Singh, A., and Mujahid, M. 2012. Performance of biofortified spring wheat genotypes in target environments for grain zinc and iron concentrations. *Field Crops Research.* 137: 261–267.

Wang, B. and Qiu, Y.L. 2006. Phylogenetic distribution and evolution of mycorrhizas in land plants. *Mycorrhiza.* 16: 299–363.

Welch, R.M. and Graham, R.D. 2005. Agriculture: the real nexus for enhancing bioavailable micronutrients in food crops. *Journal of Trace Elements in Medicine and Biology.* 18: 299–307.

White, P.J. and Broadley, M.R. 2005. Biofortifying crops with essential mineral elements. *Trends Plant Science.* 10: 586–593.

White, P.J. and Broadley, M.R. 2009. Biofortification of crops with seven mineral elements often lacking in human diets – iron, zinc, copper, calcium, magnesium, selenium and iodine. *New Phytologist.* 182: 49–84.

Zhu, C., Naqvi, S., Gomez-Galera, S., Pelacho, A.M., Capell, T., and Christou, P. 2007. Transgenic strategies for the nutritional enhancement of plants. *Trends Plant Science.* 12: 548–555.

Zimmermann, M.B. and Hurrell, R.F. 2002. Improving iron, zinc and vitamin A nutrition through plant biotechnology. *Current Opinion in Biotechnology.* 13: 142–145.

# 9 Product Formulations

*Manju Nehra, Vandana Chaudhary and Amanjyoti*

## CONTENTS

## 9.1 INTRODUCTION

Pearl millet can be processed and consumed as ingredients in diversified foods. It is called a nutri-cereal because of its high protein, fiber, mineral and fatty acids content, as well as its antioxidant properties. Also it is an alternative food for celiacs and gluten sensitive individuals (Chandrasekara et al., 2012; Saleh et al., 2013; Annor et al., 2015). In the search for gluten free products to address the issues of celiac disease, millet is positioned as a viable raw material for such products. Its flour can be utilized and transformed to products such as beverages, porridges and baked products (Taylor and Emmambux, 2008). Pearl millet is the basic staple food in the poorest countries and is used by the poorest people. For human consumption, it can be used in a variety of ways, including both leavened and unleavened breads, in porridges, and can also be boiled or steamed. It is also used as an ingredient in alcoholic beverages (Kulthe et al., 2018). Millet is an important ecological food security crop known for its drought resistance and nutritional quality and can be an immediately beneficial subsistence food for a nutrient-scarce populace. This group of cereal crops has significant potential in widening the genetic diversity of the food basket and ensuring improved food and nutrition security (Mal et al., 2010). Millet grains, before consumption and for preparation of food, are usually processed by commonly used traditional processing techniques, including decorticating, malting, fermentation, roasting, flaking and grinding to improve their edible, nutritional and sensory properties (Saleh et al., 2013).

## 9.2  TRADITIONAL FOODS AND BEVERAGES

There are a huge number of traditional millet foods and beverages. They can be categorized as wholegrain foods, foods made from meal/flour, and non-alcoholic and alcoholic beverages. These traditional products are consumed in Africa, the Indian subcontinent and East Asia. Because of the vast number of different local variations, this chapter is limited to describing representative examples of the various categories of foods and beverages.

### 9.2.1  Wholegrain Foods

Many grains, including finger millet, are popped in India. The process involves moistening the grains to about 19% moisture, allowing them to temper for several hours, then agitating the grain in a bed of hot sand (240°C) for a few minutes (Malleshi and Hadimani, 1994). Popping removes the outer pericarp. The popped grain may be consumed as a snack or further processed by milling. Unfortunately, the quality of the products is generally poor, due to rancidity and contamination with sand. In Gujarat province, India, wholegrain finger millet may be cooked to produce a rice-like product called *kichadi* (Subramanian and Jambunathan, 1980).

### 9.2.2  Foods Made from Meal/Flour

Not unnaturally, there is a wide range of traditional millet foods produced from meal (coarsely ground grain) or flour. Such foods include flatbreads, couscous, dumplings and porridges.

### 9.2.3  Flatbreads

These pancake-like breads are staples in the Horn of Africa (Ethiopia, Eritrea, and Sudan). They are made from a variety of different cereals, especially millets. A feature of many of the flatbreads is that the flour undergoes a mixed lactic acid bacteria and yeast fermentation (Gashe et al., 1982), which gives them a somewhat leavened texture and an acidic flavor. Probably the two most well known of these flatbreads are *injera* and *kisra*. *Injera*, from Ethiopia and Eritrea, is a large (approximately 50 cm diameter), spongy textured pancake about 5 mm thick. It has a honeycomb-like appearance, very similar to an English crumpet. Teff, followed by finger millet are preferred for making *injera*. *Injera* is served with just about anything, especially spicy sauces. *Kisra*, from Sudan, in contrast, is a thin, flexible wafer (1–1.5 mm thick) with neither holes nor a spongy texture (Badi et al., 1989). *Kisra* is served with stews (*mullah* or *tabbikh*), relish or sauce, or on its own seasoned with salt and chillies (Ejeta, 1982). In southern India and Sri Lanka, millets may be used to make *dosa*, a thin, fermented pancake that contains black gram (mung bean) (Murty and Kumar, 1995). In India, an unfermented pancake, called *roti*, is produced from pearl millet, small millets, sorghum or maize flour. *Roti* is also known as *chapatti* in other parts of India and East Africa, such as Tanzania. This popular staple in India is a very thin (1.3–3.0 mm), 12–25 cm diameter pancake with a soft, flexible puffed texture. *Rotis* are served with vegetables, meat, fermented milk

**FIGURE 9.1**   Pearl Millet Flour Chapatti

products, pickles, chutney, or sauce (Murty and Kumar, 1995). The preparation of *roti* entails taking about 50 g of flour, which is mixed with about 45 ml of warm water. The flour–water mixture is kneaded on a wooden board to obtain a cohesive dough. The dough is made into a ball, which is pressed by a wooden rod into a thin (about 1.3 to 3 mm thick) circular sheet, which is then baked for about one minute on an earthen or iron pan. In Ethiopia, a similar sweet, unleavened flatbread is called *kitta* and is preferably made from teff (Bultosa and Taylor, 2004). Siroha et al. (2016) reported *chapatti* making from pearl millet flour (Figure 9.1) and observed the antioxidant properties of chapatti. It was reported that antioxidant properties of chapatti decreased as compared to flour.

### 9.2.4 DUMPLINGS AND OTHER DOUGH PRODUCTS

The Pedi people of northern South Africa traditionally prepared boiled dough dumplings from wholegrain pearl millet meal (Quin, 1959). These dumplings, or breads, from pearl millet are described as being greenish-brown in color with a firm crumbly texture and a pleasant, slightly bitter, nutty, musty, sweet taste. A similar Indian dumpling product is called *mudde* (Malleshi and Hadimani, 1994). Also in India, steamed millet dough may be fried to produce a snackfood called *ponganum* (Subramanian and Jambunathan, 1980). Lin et al. (1998) describe a similar type of product from the Shanxi province in the north of China from proso millet, called an oil pudding. Strips of steamed millet dough are wrapped around cooked red beans and fried. Apparently, the oil pudding has a sweet aroma and a delicate texture on the inside.

### 9.2.5 PORRIDGES

There is an almost infinite range of traditional porridges that may be made from millets. The porridges range in consistency from stiff, like mashed potato, to a runny, spoonable gruel. The consistency is primarily related to the solids content of the porridge, which ranges from approximately 30% down to 10%. The serving temperature also plays a role in porridge consistency. Viscous hot porridges will invariably set when allowed to cool. Porridges also vary greatly in flavor. They are frequently soured by lactic acid fermentation or the addition of an acid, such as tamarind juice and, today, even mayonnaise. However, pearl millet and finger millet are still used by rural people across the region. In the Sahel region, stiff porridges are commonly called t6 and decorticated pearl millet is probably the most popular cereal used. In Mali, t6 is often made alkaline by the addition of wood or millet/sorghum stalk leachate or lime (calcium oxide) (Rooney et al., 1986). The pH of the t6 is around 8.2 and it is served cool with a sauce. In Shanxi province, China, foxtail millet porridge is a traditional food (Lin et al., 1998).

## 9.2.6 THICK PORRIDGES

The most common and simple food prepared from millets is porridge. Stiff (thick) porridges are consumed in almost all countries where sorghum and millets are cultivated (Murty and Kumar, 1995). Soft (thin) porridges are also a simple food product made from millets. The basic difference between thick and thin porridges is the concentration of flour. Generally, thick porridges are solid, and can be eaten with the hand, while thin porridges are fluid and can be drunk from a cup, or eaten with a spoon. The preparation of stiff porridge entails adding flour to boiling water in increments accompanied by vigorous stirring. The flour is cooked until it forms a thick, homogeneous and well gelatinized mass devoid of lumps. In some countries, the pH of the resulting stiff porridge will vary greatly, depending on the ingredients added (Dendy, 1995). In Burkina Faso, for example, the pH of thick porridge, called *tô*, is prepared by cooking flour in water to which tamarind extract or lemon juice is added to produce an acidic medium (Murty and Kumar, 1995). In some parts, such as Nigeria, Ghana, Zimbabwe, Tanzania, Kenya, Uganda and some parts of south India, plain water without any additives is used for making stiff porridge, which results in a neutral pH medium. In Botswana and Sudan, the flour is soured and fermented for at least 18 h before cooking, resulting in a fermented porridge. The different types of stiff porridges are prepared from millets and sorghum.

*Dalia* is a traditional breakfast cereal of north India, usually prepared from wheat. It is generally consumed by infants, young children, elderly people and health-conscious consumers. Pearl millet is also used for making *dalia* and *khichri* in some areas. Mridula et al. (2015) reported that multigrain *dalia* (MGD) formulations were prepared utilizing sprouted wheat and a mixture of three other grains (barley, sorghum and pearl millet). The proportion of grits of different sprouted grains influenced the overall quality of different MGD samples significantly. MGD showed very good overall sensory acceptability scores, rich in crude fiber, calcium and iron content, and low cooking time.

## 9.2.7 UGALI

The stiff porridge prepared from any cereal flour (or yam, cassava, or potato flour) is known as *ugali* in Eastern, and some parts of Southern, Africa (Kajuna, 1995). The most popular *ugali* in the region is made from maize flour, but other cereals, such as millets, sorghum, and such roots and tubers as yams, potatoes and cassava may be used (Bangu et al., 2000). In millets (such as finger millet), white grain types are preferred for *ugali* preparation. The flour is generally obtained by the traditional decorticating and grinding methods. Soft and brown grains are usually pounded dry and winnowed before grinding into flour. In urban areas, hammer mills may be used. The cooking procedure of *ugali* entails boiling a predetermined amount of water in a pan. Flour (from millet or any other source) is mixed with boiling water and vigorously stirred and kneaded by a paddle for about 10 min, until a thick, consistent mixture is obtained.

### 9.2.8 Non-alcoholic Beverages

A very popular pearl millet beverage in Namibia is called *oskikundu*. It is a lactic acid fermented product made from cooked pearl millet flour with added sorghum malt flour (Taylor, 2004). *Oskikundu* is greenish-brown in color with a slightly viscous consistency and a buttery, sour taste. Similar products called *togwa*, made from maize meal and finger millet malt (Oi and Kitabatake, 2003), and *kununzaki* (Ayo, 2005) (Plate 6.2), which may be made from pearl millet and white fonio, are produced in Tanzania and Nigeria, respectively. In Zimbabwe, a traditional fermented beverage combines milk and finger millet to give a highly nutritious product (Mugocha et al., 2000).

In alcoholic beverages across much of Africa, pearl millet and finger millet are still used widely, for example, to make traditional African beers. In Southern Africa, traditional beer is often called opaque beer because of its appearance, resulting from semi-suspended particles from the cereal, gelatinized starch= and yeast. Quin (1959) describes how the Pedi people made beer from 100% pearl millet malt. The beer was greenish-brown in color with a milk-like effervescent consistency and a pleasant, musty, bitter–sour taste. The lactic acid content was 1.8%. These opaque beers are effervescent because they are not pasteurized and they are consumed when they are actively fermenting. Opaque beers have a relatively low alcohol content, up to 3%. Today, in Bulawayo, Zimbabwe, pearl millet is malted on a large commercial scale in modern, pneumatic-type maltings and used as an ingredient in an industrially brewed opaque beer called *Ndlovo*, which means elephant in the Ndebele language. A similar product from the Balkans, Egypt and Turkey is *bosa*, also called *busa* or *bouza* (Arici and Daglioglu, 2002). The name is derived from *buze*, the Persian word for millet. *Boza* can be brewed from various cereals, but proso millet is preferred. It is a thick liquid, pale yellow in color, with a characteristic acid–alcoholic aroma. The alcohol content is generally low, less than 1%, but *boza* from Egypt can contain up to 7%. In Ethiopia, finger millet and teff are used as ingredients to make a traditional opaque beer called *tella* and a spirit called *katikalla* (Bultosa and Taylor, 2004). In the Himalayas, a traditional beer called *chhang* or *jnard/ jaanr* is produced from finger millet (Malleshi and Hadimani, 1994). Interestingly, the brewing process does not involve malting the grain (Basappa, 2002). In contrast, the traditional beers in West Africa, which are made from cereals including pearl millet, are substantially clear. These beers are variously known as *burukutu, dolo, pito, sulim,* or *talla* (Taylor and Belton, 2002). They are characterized by being filtered, but remain somewhat cloudy. They are sweetish with a slightly sour taste and a fruity aroma, and contain 1–5% alcohol (Demuyakor and Ohta, 1993).

## 9.3   STEAM COOKED PRODUCTS

### 9.3.1   KUDUMU

In India, fermented sorghum-millet flour is cooked over steam to make a product called *kudumu*. Wet milled sorghum-millet batter can be mixed with wet milled black gram batter, fermented overnight and poured into small cakes or moulds which are steam cooked to give a product called *idli*. *Idli* is soft, moist and spongy, with a slightly sour taste (Murty and Kumar, 1995).

### 9.3.2   FURAH

In Northern Nigeria and Niger, a traditional pearl millet product called *furah* is popular and commonly sold in the markets. Slightly fermented and humid flour from millet is shaped into balls, which are placed in a small quantity of water and steam-cooked for about one hour, after which they are pounded in a mortar to give a highly viscous paste. The paste is then rolled in fine millet flour and sold as *furah* (Murty and Kumar, 1995).

## 9.4   LADOO

Singh and Sehgal (2008) prepared *ladoo* from pearl millet grain and *ladoo* was evaluated for chemical composition, *in vitro* protein and starch digestibility. Roasted and dehusked chickpea, groundnut and jiggery were used for preparation of *ladoo*. Two types of *ladoo* were developed: type I from 50% popped pearl millet with roasted and dehulled chickpea and groundnut, and type II from 100% popped pearl millet. It was observed that type I popped pearl millet *ladoo* had significantly higher calcium, phosphorus and iron content. Higher polyphenol and phytic acid and lower *in vitro* protein and starch digestibility were also found in type I *ladoo*.

## 9.5   BAKERY PRODUCTS

Baking is considered a simultaneous heat and mass transfer process, characterized by a rapid increase of the core temperature and the development of a dry surface crust. In consequence, baking process conditions—oven temperature, baking time and oven humidity—strongly influence the development of all quality attributes (Hadiyanto et al., 2007). From the different quality attributes, surface crust color is one of the critical characteristics, since its value directly affects initial acceptance by consumers (Mundt and Wedzicha, 2007).

Biscuits are the most popular bakery items, consumed by nearly everyone (Figure 9.2). This is mainly due to their ready to eat nature, good nutritional quality, availability in numerous varieties and affordability. Based on production statistics, the top three producers of biscuits are the United States, China, and India, respectively (Misra and Tiwari, 2014). Among the ready-to-eat products, biscuits are of significant importance because they are widely accepted, affordable and they have a relatively long shelf life (Florence-Suma et al., 2012). Adebiyi et al. (2016) investigated the effect of fermentation and malting on flour and biscuits prepared from pearl millet flour. Malting and fermentation were observed to result in a degradation of protein and starch polymers to other smaller constituent structures. This reflected changes in the microstructure and physicochemical properties of the pearl millet flour and consequent baked biscuits. These changes resulted in enhanced biscuit hardness and functionalities for potential application in baked food products. Adebiyi et al. (2017) prepared biscuits from malted and fermented pearl millet flour. In this study, fermentation and malting were found to increase the nutritional value, mineral composition and amino acid content of both the flours and the resultant biscuits. The malting process also modified the health beneficent polyphenols. Sensory properties revealed that biscuits prepared from malted pearl millet flour showed the highest overall acceptance. Total phenolic content and total flavanoid content of malted flour was found to be higher as compared to the control and fermented samples. Obafaye and Omoba (2018) reported utilization of orange peel flour for the preparation of pearl millet biscuits. The study investigated the effect of orange peel powder at different levels of substitution (5%, 10%, 15% and 20%) in the production of pearl millet biscuit on the proximate, antioxidant potentials, dietary fiber and consumer acceptance. It was observed that the antioxidant activity of biscuits increased with the incorporation of orange peel powder.

The word cookie refers to small cakes, derived from the Dutch words *kockje* or *koekie*. Cookies mainly consist of the same ingredients as cake,s except that they have a lower proportion of liquid and a higher proportion of sugar and fat to flour (Figure 9.3). Cookies represent the largest category of

**FIGURE 9.2**    Biscuits prepared from pearl millet flour

**FIGURE 9.3** Cookies prepared from pearl millet

snacks in the bakery industry and can serve as an effective vehicle for the supply of nutrients to consumers. They represent the baked product containing three major ingredients: flour, fat and sugar. Cookies have low water content (1–5%) and can also contain minor ingredients such as leavening agents, salt, emulsifiers and yeast (Pareyt and Delcour, 2008). Awolu et al. (2017) studied the effect of the addition of pearl millet, subjected to different processing techniques, to the composite flour comprising rice, soybeans and tigernut on the proximate composition, with the functional and pasting properties evaluated showing that the addition of fermented pearl millet flour was the best compared to debranned and malted pearl millet flours. Fermentation increases the protein content of the composite flours; malting, fermentation and debranning increases the functional properties. In general, all the composite flours had good nutritional and flour quality which could be used in producing acceptable cookies.

Bread is an important staple food worldwide, baked generally from wheat flour, sugar, fat, salt and leavening agent (Figure 9.4). Wheat is the main crop for breadmaking due to its baking performance related to its gluten protein content (Dewettnick et al., 2008). With the constant search for diversity and innovation in foods, an alternative market for nutritious baked products has emerged to satisfy the interest of health-conscious people in their diet (Ronda et al., 2015). Generally, bread is mainly prepared from wheat flour. To improve the nutritional profile of wheat bread, the substitution of wheat flour by other flours is used. Maktouf et al. (2016) reported bread preparation using pearl millet and wheat flours. It was observed that the addition of 5% of millet flour was able to improve the rheological properties of the dough as well as the specific volume and texture of the bread.

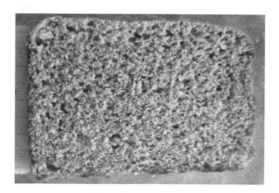

**FIGURE 9.4** Bread prepared from debranned pearl millet flour

## 9.6 PASTA

Pasta is one among the ready-to-cook cereal foods that comprises spaghetti, noodles, vermicelli, etc. The popularity of pasta has swelled enormously in the Indian market due to easy convenience and palatability. Rathi et al. (2004) studied the influence of depigmentation of pearl millet (*Pennisetum glaucum* L.) on sensory attributes, nutrient composition, *in vitro* protein and starch digestibility of pasta. Pasta products were prepared from semolina, unprocessed pearl millet (T-I) and depigmented pearl millet (T-II). The protein, fat, ash and dietary fiber contents of pearl millet-based pastas (T-I and T-II) were significantly ($P < 0.05$) higher than those of the control pasta. Results indicated that depigmentation of pearl millet significantly ($P < 0.05$) improved the sensory attributes, especially the color of the pasta. Jalgaonkar et al. (2018) prepared pasta from a blend of wheat semolina and pearl millet flour (50:50), having supplemented it with defatted soy flour (DSF) (5%, 15% and 25%), carrot powder (CP) (5%, 10% and 15%), mango peel powder (MPP) (5%, 10% and 15%), and moringa leaves powder (MLP). Maximum incorporation of 15% DSF, 10% CP, 5% MPP and 3% MLP was found suitable in terms of color, cooking loss (<8%), hardness, and sensory quality.

## REFERENCES

Adebiyi, J. A., Obadina, A. O., Adebo, O. A. and Kayitesi, E. 2017. Comparison of nutritional quality and sensory acceptability of biscuits obtained from native, fermented, and malted pearl millet (Pennisetum glaucum) flour. *Food Chemistry* 232: 210–217.

Adebiyi, J. A., Obadina, A. O., Mulaba-Bafubiandi, A. F., Adebo, O. A. and Kayitesi, E. 2016. Effect of fermentation and malting on the microstructure and selected physicochemical properties of pearl millet (Pennisetum glaucum) flour and biscuit. *Journal of Cereal Science 70*: 132–139.

Annor, G. A., Marcone, M., Corredig, M., Bertoft, E. and Seetharaman, K. 2015. Effects of the amount and type of fatty acids present in millets on their in vitro starch digestibility and expected glycemic index (eGI). *Journal of Cereal Science* 64: 76–81.

Arici, M. and Daglioglu, O. 2002. Boza: A lactic fermented cereal beverage as a traditional Turkish food. *Food Reviews International* 18: 39–48.

Awolu, O. O., Olarewaju, O. A. and Akinade, A. O. 2017. Effect of the addition of pearl millet flour subjected to different processing on the antioxidants, nutritional, pasting characteristics and cookies quality of rice-based composite flour. *Journal of Nutritional Health and Food Engineering* 7(2): 00232.

Ayo, J. A. 2005. Effect of acha (*Digitaria exilis Staph*) and millet (*Pennisetum typhodium*) grain on kunun zaki. *British Food Journal* 106: 512–519.

Badi, S. M., Bureng, P. L. and Monowar, L. Y. 1989. Commercial production: A breakthrough in kisra technology. In Dendy, D. A. V. ed. *ICC 4th Quadrennial Symposium on Sorghum and Millets, Lausanne, Switzerland*, Vienna: International Association for Cereal Science and Technology, pp. 31–45.

Bangu, N. T. A., Kajuna, S. T. A. R. and Mittal, G. S. 2000. Storage and loss moduli for various stiff porridges. *International Journal of Food Properties* 3(2): 275–282.

Basappa, S. C. 2002. Investigations on chhang from finger millet (Eleucine coracana Gaertn.) and its commercial prospects. *Indian Food Industry* 21: 46–51.

Bultosa, G. and Taylor, J. R. N. 2004. Teff. In Wrigley, C., Corke, H., and Walker, C. E. eds. *Encyclopedia of Grain Science*, Vol 3. Amsterdam: Elsevier, pp. 281–290.

Chandrasekara, A., Naczk, M. and Shahidi, F. 2012. *Effect of processing on the Chennai*, India: M.S. Swaminathan Research Foundation: 1–185.

Demuyakor, B. and Ohta, Y. 1993. Characteristics of single and mixed culture fermentation of pito beer. *Journal of Science Food and Agricultural* 62: 401–408.

Dendy, D. A. V. 1995. ed. *Sorghum and Millets: Chemistry and Technology*, St. Paul, MN: AACC Intl.

Dewettnick, K., VanBockstaele, F., Kühne, B., VandeWalle, D., Courtens, T. M. and Gellynck, X. 2008. Nutritional value of bread: influence of processing, food interaction and consumer perception. *Journal of Cereal Science* 48: 243–257.

Ejeta, G. 1982. Kisra quality: Testing new sorghum varieties and hybrids. In Mertin, J. V. ed. *Proceedings of the International Symposium on Sorghum Grain Quality*, Patancheru, India: ICRISAT, pp. 67–78.

Florence-Suma, P., Urooj, A., Asha, M. R. and Rajiv, J. 2012. Sensory, physical and nutritional qualities of cookies prepared from pearl millet (*Pennisetum typhoideum*). *Journal of Food Processing and Technology* 5: 377 52–56. doi:10.4172/2157-7110.1000377.

Gashe, B. A., Girma, M. and Bisrat, A. 1982. Tef fermentation. I. The role of microorganisms in fermentation and their effect on the nitrogen content of tef. SINET. *Ethiopian Journal of Science* 5: 69–76.

Hadiyanto, H., Asselman, A., van Straten, G., Boom, R. M., Esveld, D. C. and van Boxtel, A. J. B. 2007. Quality prediction of bakery products in the initial phase of process design. *Innovative Food Science and Emerging Technologies* 8: 285–298.

Jalgaonkar, K., Jha, S. K. and Mahawar, M. K. 2018. Influence of incorporating defatted soy flour, carrot powder, mango peel powder, and moringa leaves powder on quality characteristics of wheat semolina-pearl millet pasta. *Journal of Food Processing and Preservation 42*(4): e13575.

Kajuna, S. T. A. R. 1995. Low cost technology for hulling maize. Agricultural mechanization in Asia. *Africa and Latin America* 26(3): 39–40, 48.

Kulthe, A. A., Thorat, S. S. and Khapre, A. P. 2018. Nutritional and sensory characteristics of cookies prepared from pearl millet flour. *The Pharma Innovation Journal* 7 (4): 908–913.

Lin, R., Li, W. and Corke, H. 1998. Spotlight on Shanxi province China: Its minor crops and specialty foods. *Cereal Foods Worm* 43: 189–192.

Maktouf, S., Jeddou, K. B., Moulis, C., Hajji, H., Remaud-Simeon, M. and Ellouz-Ghorbel, R. 2016. Evaluation of dough rheological properties and bread texture of pearl millet-wheat flour mix. *Journal of Food Science and Technology* 53(4): 2061–2066.

Mal, B., Padulosi, S. and Ravi, S. B. 2010. Minor millets in South Asia: Learnings from quality, processing, and potential health benefits. *Comprehensive Reviews in Food Science and Food Safety* 12(3): 281–295.

Malleshi, N. G. and Hadimani, N. A. 1994. Nutritional and technological characteristics of small millets and preparation of value-added products from them. In Riley, K. W., Gupta, S. C., Seetharman, A., and Mushonga, J. N. eds. *Advances in Small Millets*, New York: International Science Publisher, pp. 271–287.

Misra, N. N. and Tiwari, B. K. 2014. Biscuits. In W. Zhou, Y. H. Hui, I. De Leyn, M. A. Pagani, C. M. Rosell, J. D. Selman, and N. Therdthai. *Bakery Products Science and Technology* Second Edition, John Wiley & Sons, Ltd, 585–601.

Mridula, D., Sharma, M. and Gupta, R. K. 2015. Development of quick cooking multi-grain dalia utilizing sprouted grains. *Journal of Food Science and Technology* 52(9): 5826–5833.

Mugocha, P. T., Taylor, J. R. N. and Bester, B. H. 2000. Fermentation of a composite finger millet–dairy beverage. *World Journal of Microbiology Biotechnology* 16: 341–344.

Mundt, S. and Wedzicha, B. L. 2007. A kinetic model for browning in the baking of biscuits: effects of water activity and temperature. *LWT-Food Science and Technology* 40(6): 1078–1082.

Murty, D. S. and Kumar, K. A. 1995. Traditional uses of sorghum and millets. In Dendy, D. A. V. ed. *Sorghum and Millets: Chemistry and Technology*, St. Paul, MN: American Association of Cereal Chemists, pp. 185–222.

Obafaye, R. O. and Omoba, O. S. 2018. Orange peel flour: A potential source of antioxidant and dietary fiber in pearl-millet biscuit. *Journal of Food Biochemistry* 42(4): e12523.

Oi, Y. and Kitabatake, N. 2003. Chemical composition of East African traditional beverage, togwa. *Journal of Agricultural Food Chemistry* 51: 7024–7028.

Pareyt, B. and Delcour, J. A. 2008. The role of wheat flour constituents, sugar and fat in low moisture cereal based products: a review on sugar-snap cookies. *Critical Reviews in Food Science and Nutrition* 48(9): 824–839.

Quin, P. J. 1959. *Foods and the Feeding Habits of the Pedi*, Johannesburg: Witwatersrand University Press.

Rathi, A., Kawatra, A. and Sehgal, S. 2004. Influence of depigmentation of pearl millet (*Pennisetum glaucum* L.) on sensory attributes, nutrient composition, in vitro protein and starch digestibility of pasta. *Food Chemistry* 85(2): 275–280.

Ronda, F., Abebe W., Perez-Quirce S and Collar C. 2015. Suitability of tef varieties in mixed wheat flour bread matrices: A physico-chemical and nutritional approach. *Journal of Cereal Science* 64: 139–146.

Rooney, L. W., Kirleis, A. W. and Murty, D. S. 1986. Traditional foods from sorghum. In Pomeranz, Y. ed. *Advances in Cereal Science and Technology*, Vol. VIII, St. Paul, MN: American Association of Cereal Chemists, pp. 317–353.

Saleh, A. S., Zhang, Q., Chen, J. and Shen, Q. 2013. Millet grains: nutritional quality, processing, and potential health benefits. *Comprehensive Reviews in Food Science and Food Safety* 12(3): 281–295.

Singh, G. and Sehgal, S. 2008. Nutritional evaluation of ladoo prepared from popped pearl millet. *Nutrition and Food Science* 38(4): 310–315.

Siroha, A. K., Sandhu, K. S. and Kaur, M. 2016. Physicochemical, functional and anti-oxidant properties of flour from pearl millet varieties grown in India. *Journal of Food Measurement and Characterization* 10: 311–318.

Subramanian, V. and Jambunathan, R. 1980. Traditional methods of processing sorghum (*Sorghum bicolor*) and pearl millet (*Pennisetum americanum*) grains in India. In Dendy, D. A. V. ed. *International Association for Cereal Chemistry, 10th Congress, Symposium: Sorghum and Millet Processing*, Vienna: ICC, pp. 115–118.

Taylor, J. R. N. 2004. Foods and nonalcoholic beverages. In Wrigley, C., Corke, H., and Walker, C. E. eds. *Encyclopedia of Grain Science*, Vol. 1, Amsterdam: Elsevier, pp. 380–390.

Taylor, J. R. N. and Belton, P. S. 2002. Sorghum. In Belton, P. S. and Taylor, J. R. N. eds. *Pseudocereals and Less Common Cereals*, Berlin: Springer-Verlag, pp. 25–91.

Taylor, J. R. N., Emmambux, M. N. 2008. Gluten-free cereal products and beverages. In Arendt, E. K., Bello, F. D. eds. *Gluten-free Foods and Beverages from Millets*, Amsterdam: Elsevier Inc., pp. 119–148.

# Index